Breeding Zinc Crops for Better Human Health

Mahalingam Govindaraj
Velu Govindan • Natalia Palacios
Editors

Breeding Zinc Crops for Better Human Health

Springer

Editors
Mahalingam Govindaraj
Alliance of Bioversity International and the
International Center for Tropical
Agriculture (CIAT)
Palmira, VALLE, Colombia

Velu Govindan
Global Wheat Program
International Maize and Wheat
Improvement Center (CIMMYT)
Texcoco, Estado de México, Mexico

Natalia Palacios
International Maize and Wheat
Improvement Center (CIMMYT)
Texcoco, Estado de México, Mexico

ISBN 978-3-031-84341-9   ISBN 978-3-031-84342-6   (eBook)
https://doi.org/10.1007/978-3-031-84342-6

© The Editor(s) (if applicable) and The Author(s) 2025. This book is an open access publication.

Open Access This book is licensed under the terms of the Creative Commons Attribution 4.0 International License (http://creativecommons.org/licenses/by/4.0/), which permits use, sharing, adaptation, distribution and reproduction in any medium or format, as long as you give appropriate credit to the original author(s) and the source, provide a link to the Creative Commons license and indicate if changes were made.
The images or other third party material in this book are included in the book's Creative Commons license, unless indicated otherwise in a credit line to the material. If material is not included in the book's Creative Commons license and your intended use is not permitted by statutory regulation or exceeds the permitted use, you will need to obtain permission directly from the copyright holder.
The use of general descriptive names, registered names, trademarks, service marks, etc. in this publication does not imply, even in the absence of a specific statement, that such names are exempt from the relevant protective laws and regulations and therefore free for general use.
The publisher, the authors and the editors are safe to assume that the advice and information in this book are believed to be true and accurate at the date of publication. Neither the publisher nor the authors or the editors give a warranty, expressed or implied, with respect to the material contained herein or for any errors or omissions that may have been made. The publisher remains neutral with regard to jurisdictional claims in published maps and institutional affiliations.

This Springer imprint is published by the registered company Springer Nature Switzerland AG
The registered company address is: Gewerbestrasse 11, 6330 Cham, Switzerland

If disposing of this product, please recycle the paper.

Foreword

The poor and rich in Low- and Middle- Income Countries (LMICs) eat food staples in about the same large amounts. It is well known that food staples, eaten day in and day out at every meal, provide a low-cost source of energy which staves off hunger for families who desire but cannot afford foods that are more dense in minerals and vitamins such as vegetables, fruits, legumes, and animal products. However, especially for the poor, it is not widely recognized that food staples provide a significant base of selected minerals and vitamins that are essential for health, and in particular zinc.

Food staple plant breeders face multiple changes. Global population continues to grow and production is threatened by climate change. Productivity, i.e., yields, must increase to keep food staple prices low for the poor. Yet at the same time, increased yields have led to long-run declines in mineral and vitamin densities in food staples, and increased atmospheric CO_2 levels due to climate change threaten to reduce these densities even further.

Hidden hunger remains a serious public health problem. Getting plant breeders to reverse these trends—that is, to increase the densities of minerals and vitamins in food staples—now is a proven and cost-effective strategy to help in the fight against hidden hunger, particularly for the poor. When HarvestPlus first started 20 years ago, one of the technical challenges was to develop low-cost and high-throughput methods to measure the levels of trace minerals of seeds in breeding programs. There was skepticism both that high yields could be combined with zinc density and that zinc in biofortified crops would be absorbed in sufficient amounts to have a public health impact. Moreover, would farmers adopt the biofortified varieties and would the public buy and consume them?

These issues are all addressed in the positive in the chapters in this volume, which focus on zinc wheat for India and Pakistan, but also touch on zinc rice for Bangladesh, and zinc maize for the tropics in general. A final chapter addresses the complementary approach of adding zinc to fertilizers and foliar sprays, an approach referred to as "agronomic biofortification."

The contributors to this book have expertly woven together the science, technology, and prospects for breeding zinc-rich crops, making it a crucial read for plant breeders, researchers, agronomists, and policymakers alike. I hope that the insights shared in this book will inspire new collaborations and discoveries that will lead to the next generation of staple crop varieties that can help nourish the planet.

World Food Prize Laureate -2016, Founding Director Howarth E. Bouis
HarvestPlus, a Challenge Program of the CGIAR
Emeritus Fellow, International Food Policy Research Institute
Washington, DC, USA

Foreword

In the face of global challenges like food insecurity, malnutrition, and a growing population, the importance of enhancing the nutritional quality of crops has never been more pressing. Zinc, an essential micronutrient, plays a crucial role in human health, influencing immunity, growth, and development. However, millions of people worldwide suffer from zinc deficiency, particularly in developing countries where staple crops often lack sufficient zinc levels. This deficiency contributes to various health problems, including stunted growth, weakened immune systems, and cognitive impairments.

To address this issue, scientists and agricultural experts are focusing on breeding zinc-rich crops as a sustainable and scalable solution. By enhancing the zinc content of staple foods like rice, wheat, and maize, it is possible to improve the dietary intake of this vital nutrient, especially for populations that rely heavily on these crops for sustenance.

This book represents a combination of scientific innovation, agricultural advancements, and nutrition-focused research aimed at creating crops that not only provide caloric energy but also deliver the nutrients necessary for optimal health. Breeding zinc-rich crops holds immense promise for improving public health, particularly in regions where malnutrition remains a significant barrier to socioeconomic development. For instance, wheat is a major source and an ideal vehicle for delivering increased quantities of zinc, iron, and other valuable bioactive compounds to populations that consume it as a staple crop. For that reason, International Maize and Wheat Improvement Center (Centro Internacional de Mejoramiento de Maíz y Trigo) (CIMMYT) has been mainstreaming grain zinc across its breeding pipelines to achieve enhanced zinc content in most elite wheat lines distributed globally.

Congratulations to the editors and contributing authors for promoting biofortification breeding and mainstreaming nutrients in staple crops. The chapters in this book delve into the strategies, challenges, and breakthroughs involved in the development and dissemination of zinc-enriched crops. Through the integration of biotechnology, conventional breeding, and agronomic practices, scientists are working to create crops that can contribute to address global zinc deficiency and significantly improve nutrition for present and future generations.

International Maize and Wheat Improvement
Center (CIMMYT)
Texcoco, Estado de México, Mexico

Bram Govaerts

Contents

**High Throughput Phenotyping for Grain Zinc: Sampling
and Analytical Overview** 1
Mahalingam Govindaraj, Mahesh Pujar, and Venkatram Pravalika

Breeding and Deployment of High Zn Wheat in South Asia 17
Velu Govindan, Arun Kumar Joshi, Pradeep Bhati,
and Karthikeyan Thiyagarajan

Breeding and Deploying High-Zinc Maize in the Tropics 41
Prasanna Boddupalli, Natalia Palacios-Rojas, Felix San Vicente,
Thanda Dhliwayo, Abebe Menkir, Thokozile Ndhlela, Sudha K. Nair,
and Xuecai Zhang

**Zinc Wheat Variety Release, Seed Production, and Scaling
Up Strategies in India** 65
Chandra Nath Mishra, Amit Sharma, Satish Kumar, Disha Kamboj,
Gyanendra Singh, Arun Kumar Joshi, and Gyanendra Pratap Singh

**Zinc Wheat Variety Release, Seed Production and Scaling
Up Strategies in Pakistan** 81
Muhammad Imtiaz

Current Status of Zinc-Biofortified Rice Cultivation in Bangladesh 95
Khondoker Abdul Mottaleb, Alvaro Durand-Morat, Fazleen Abdul Fatah,
and Md. Abdur Rouf Sarkar

**Biofortified Cereals Increase Dietary Zinc Intake: Wheat
and Maize as Case Studies** 123
Swarnim Gupta and Nicola M. Lowe

**Agronomic Biofortification of Crops with Zinc: A Comprehensive
Overview** 153
Raheela Rehman, Muhammad Moaz Latif, Muhammad Ahsan Khan,
and Zaheer Ahmed

High Throughput Phenotyping for Grain Zinc: Sampling and Analytical Overview

Mahalingam Govindaraj, Mahesh Pujar, and Venkatram Pravalika

Introduction

Globally, zinc (Zn) deficiency is more prevalent than iron and vitamin A deficiency, especially in low-income countries where diets are dominated by cereals that are traditionally low in Zn content. Zinc deficiency is highly reported in children under 5 years old, in contrast to anaemia (caused by Fe deficiency) which is more common in adolescent girls and women. Zn is an essential nutrient for growth and recovery, so deficiencies can result in stunted growth, increase susceptibility to disease and infection, lengthen recovery time or, in some cases, impair recovery. They can also reduce mental capacity, increase the prevalence of maternal, neonatal, and child complications, and increase recovery times (Prasad, 2013). Strategies like dietary diversification and food supplements help in an urbanized area that requires recurrent costs and limited compliance remains challenging. Food fortification can also contribute in similar way (De et al., 2002)—sustainable in developed countries over the past 80 years but are not feasible in developing countries due to the lack of technical, operational, and financial feasibility. Therefore, fortification projects are uncommon and typically abandoned after failing tests (Dary & Mora, 2002). Biofortification, a strategy that enhance inherently higher nutrients (Fe, Zn, and Vitamin-A) into food crops through crop breeding approaches, offers a comparatively affordable, long-term, sustainable method of delivering more micronutrients, holding great promise for enhancing the nutritional status and health of

---

M. Govindaraj (✉) · V. Pravalika
Alliance of Bioversity International and the International Center for Tropical Agriculture (CIAT), Palmira, VALLE, Colombia
e-mail: m.govindaraj@cgiar.org

M. Pujar
International Crops Research Institute for the Semi-Arid Tropics (ICRISAT), Patancheru, Hyderabad, Telangana, India

undernourished populations in both rural (by growing and consuming) and urban (surplus marketed from farmers) areas of developing countries (Bouis, 2003). This strategy necessitates cross-disciplinary research into the soil, crop, and environmental factors impacting the improvement of the micronutrient concentration in crop varieties' valuable components. A vast number of potential germplasm and advanced breeding lines must be initially screened as part of the targeted biofortification breeding programme and being used in hybridization and varietal development and release. On parallel, physiological research is required to understand how micronutrients are transported from the soil to the final edible part of the crop (seeds, grains, or beans) during various stages of plant growth, how they do this via the roots, xylem, and phloem, and how they are stored in various storage tissues. Three primary strategies have been frequently employed in the context of the biofortification strategy: conventional crop breeding, transgenic biotechnology, and agronomic techniques with the optimal physical administration of nutrients through fertilizers into the soil. The biofortification strategy thus takes advantage of the favourable relationship between source (soil) and sink (seed), and as a result, any efforts or approaches that could increase root growth and, as a result, high uptake of zinc from the soil to the plants and, ultimately, edible parts like seeds, should be viewed and addressed critically in biofortification (Zuo & Zhang, 2009).

Most of the micronutrients in crop plants are not a visible trait to select visually. Thus, an essential component of a successful biofortification programme is the analysis of the Zn micronutrient found in the edible parts of crops such as seeds, grains, and beans that are intended to be biofortified. Inductively Coupled Plasma Optical Emission Spectroscopy (ICP-OES), Inductively Coupled Plasma Mass Spectrometry (ICP-MS), Atomic Absorption Spectrophotometer (AAS), and colorimetry (Staining technique) are a few of the well-established techniques and tools that are available for Zn analysis in plant samples. These were mostly requiring trained skill sets, chemicals, and destructive methods. However, more advanced analytical tools with great accuracy, efficiency, and throughput are necessary to accelerate progenies selection in breeding programme. Energy Dispersive X-ray fluorescence (XRF) spectroscopy provides high throughput analyses with variety of attractive features, including the ability to analyse many elements in a single run with little to no sample preparation (non—destructive) and the ability to the detection limit of a parts per million (ppm) levels. Near-infrared spectroscopy (NIRS) is another such high throughput, non-destructive instrument. However, the correlation between XRF results and ICP results are very significant and positive than NIRS (Rai et al., 2012; Govindaraj et al. 2016a, b). Therefore, this chapter provides insights on high throughput XRF analytical techniques, Zn micronutrient analysis, sample preparation, and their revolutionary contributions to the accelerated biofortification programme across a variety of staple food crops over a decade.

Grain Sampling and General Precautions

Grain sampling is a crucial stage since it is a raw material that determines the quality of the grain's value and composition attributes of a particular variety or hybrid. While breeding for Zn, gathering a representative sample (Fig. 1), grain composition varies according to crop germplasm, grown environmental conditions, physiological age (maturity and storage durations) and position of panicles grains were sampled. Therefore, it recommends having representative samples by limiting soil and dust contamination from harvesting or postharvest processing equipment constitutes a significant analytical difficulty. Dirt or dust on the user's hands or tools, applying skincare items to bare hands, such as hand lotions, soiled or corroded components, crops that lodge, causing the inflorescence to lie in the ground or water (in the case of cereals) and inadvertent mixing with other samples or touch are common sources of contamination found while sampling or processing crops in the field or a lab. Therefore, the personnel who gather and handle samples must be aware of the significance of preventing contamination. One must follow good practices like clearing the area of dust and plant matter using a clean hand glove, brush, a clean cloth, or compressed air before and after using the equipment. While working with a variety of types or samples at once, cross-contamination can be avoided by cleaning the apparatus after each sample is analysed. Additionally, to shield it from moisture and dust when not in use, the equipment should always be covered. For more sampling errors and precautions for Zn estimation, special care must be taken in the field as reported by Stangoulis and Sison (2008). The representative sample needs to be packaged with its tracking identity (labels) and should be stored appropriately until it is being used for the analysis without contamination. Sample label should have the name of the sampled crop, variety name, location, and date upon the sample collection/arrival. Brown envelopes are the ideal size for storing most dry samples between preparation and analysis. Naphthalene mothballs can be used to prevent insect infestation during long-term storage which has no significant impact Zn estimation provided the mothballs are around the sample bags but not inside or directly on the crop samples.

With regards to released biofortified cultivars sampling, the purpose is to evaluate, and monitor select target nutrients (e.g. Zn) in the real-world scenario. To obtain confident estimations of the content of all analytes of interest for a biofortified and control crop and create a robust and trusted nutrient database of biofortified crops,

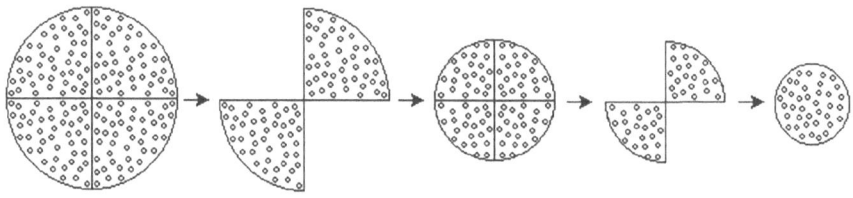

Fig. 1 Demonstration of quartering grains for representative sample of a test variety

key concepts must be understood during the sampling process. These also include (i) samples should be representative of the population, considering the expected and accepted sources of variation for each analyte in the crop; (ii) contamination can occur during processing, handling and storage of samples; (iii) degradation of analytes can occur during processing, handling and storage, and proper precautions must be taken to minimize the impact; and (iv) proper documentation of all relevant sample information is important for seasonal and longitudinal comparisons. These details should be recorded in the sampling plan and sample inventory form.

Biofortification Crop Sampling Methods

In general, most samples are collected at two different levels: sampling in the field and sampling in a storage/lab. In this section, detailed sampling protocols are discussed for five important staple food crops rice, wheat, maize, pearl millet and beans (HarvestPlus, 2008).

Sampling in the Field

At the first level, field sampling procedure for all the crops remains almost similar wherein the field team that actively takes part in the sampling is made aware of all the precautionary measures that need to be followed to avoid contamination. In addition, crops are harvested at physiological maturity preferably manually. Try to avoid sampling from sick plants (due to diseases/pest) and corner/border row plants during the sampling. The protocols for the harvesting of crops are given below:

Rice: Panicle harvesting must be done manually and must only be done on upright plants (no lodging) and still 120 days after planting/maturity stage.

Pearl millet: Obtain a sample that is truly representative of the population by hand pollinating 20 plants from the same entry with bulk pollen taken from 50–60 plants. Cover panicles with fresh paper bags as soon as they appear to lessen exposure to dust. For hybrid, cover the panicles after anthesis with paper bags. Harvest sib-mated panicles (with paper bags on) of a variety and covered panicles of a hybrid and put them in fresh, new brown paper bags once sib-mated plants have attained physiological maturity (85–90 days after planting). Open-pollinated grain sample provide unbiased Zn estimates in pearl millet while selfed samples give overestimates owing to poor grain set (Rai et al., 2013).

Maize: To prevent soil or dust contamination with iron before harvest, wrap the ears in clean paper bags. Harvest sufficient cobs to provide a representative sample once the crop reaches physiological maturity (80–120 days after planting). Harvest 100 randomly chosen ears from open-pollinated types and thresh them ear-to-row to create a balanced bulk whereas, harvest 3–10 outstanding ears and bulk thresh them for fixed inbred lines.

Wheat: When harvesting grains from a small plot, take a representative sample of roughly 10–20 randomly chosen heads and weigh it at 250 g before the main harvest. To protect the samples from contamination by dirt and soil, place them in fresh, clean, properly labelled paper envelopes.

Beans: Gather roughly 10 fully filled pods at random before the main harvest. Fill fresh, clean paper envelopes with pods as it will help to avoid contamination from dust and dirt during uprooting plants and while bulk threshing.

Sampling in the Laboratory

Rice: Thresh the panicles manually, and then put the grains in brown paper bags that are clean, unopened, and neatly labelled. These unprocessed grains are dried until their moisture content is between 12% and 14%, and they can be stored for 2 months in sterile sacks or bags. Collect a sample of rough rice weighing 50–120 g, depending on the type of mill being used. Use the Satake THU-35A dehuller to dehull the grains after replacing the contaminated component with a non-contaminating PVC compound. Additionally, the modified Suzuki rice dehuller and polisher made in Brazil, the Satake TM-05 pearling (abrasive) mill, the Grainman No. 60 (Grain Machinery Mfg. Corp., Hialeah, FL, USA), the McGill No. 2 friction-type mill, and the modified Kett mill, can be used to mill large samples (>70 g) of brown rice. Milled rice samples are packed in fresh, clean, suitably labelled sample bags and store until the analysis is complete in a cool (20 °C), dry, insect-free environment. Obtain 5 g for XRF scan.

Pearl millet: Dry the panicles in their paper/cloth bags in a clean dry sunlight area. To extract the grains from the panicles, use a thresher (non-metallic). Manually remove any remaining glumes over the grains, if any. Collect a 100-g sample of a representative grain. As soon as samples are prepared for analysis, place them in lab standards paper envelopes that are brand-new, clean, and properly labelled before storing them in dry and insect-free rooms. Scoop 5 g for XRF scan.

Maize: Using a fine-quality stainless steel knife or a plastic stick, physically remove the husk from each ear. The husked ears should be kept in a tidy basket with a loose lid. The husked ears should be placed on tidy plastic drying trays. Dry samples in an un-corroded oven at 40 °C for 5 days. Then shell the husked ears and place the kernels on a clean plastic tray and stir them well. A typical sample of about 250 g is easily taken. Kernal is usually larger (more X-ay errors) hence use a non-contaminating mill to grits/grind the kernels and transfer to new, clean, and properly labelled paper envelopes until they are ready for analysis. Obtain a test sample of 5 g (grits) for an XRF scan.

Wheat: Thresh ear heads by hand, and store the grains in a fresh and clean labelled sample bag. Use air cleaner if impurities are observed. Dry the grains in a clean oven at 75 °C for 48 h to kill the spores if the sample locations are Karnal bunt outbreak. About 5 g of whole-grain samples are sufficient for XRF analysis.

Beans: Hand-thresh the pods and get 10 g of beans. Using a cloth that has been soaked with ultra-pure water, clean the beans. Dry seeds in a hot air oven at 60 °C to a moisture level of 7–8% (~12% in fresh beans). Beans vary in size and required grits or flour to use XRF for price results. Use the grinding special bean mill for grounding. The ground samples packed in fresh, labelled bags or plastic screw-top tubes, and kept in a dry, insect-free environment until the analysis is complete. About 5-g sample is adequate for XRF scan.

Zinc (Zn) Phenotyping Methods

Several analytical tools and methods used for measuring the plant sample Zn contents in the past. That includes staining methods, AAS, ICP, NIRS. Recently novel tools such as XRF accomplished the Zn estimation in a very rapid approach. All these methods were briefly described hereunder.

Staining: The 1,5-diphenyl-thio-carbazone compound is used to stain Zn in seeds and seedlings (methanol dissolved in one litre with 300 mg) (Velu et al., 2008). The 1,5-diphenyl-thio-carbazone is dissolved in pure methanol to create the DTZ solution, which is used to detect Zn by giving the tissue part a red colour pigment. The emergence of the purple colour is the proof presence of Zn in the sample. Furthermore, the concentration of Zn in tissue samples is determined by the intensity of the colour appearance. The samples are then given scores on a scale of 1 to 5, with 1 representing no to very little colour development, 2 denoting less intense colour, 3 denoting medium colours, 4 denoting intense colour, and 5 denoting very intense colour. The corresponding staining methods were standardized to screen Zn in the breeding lines in pearl millet (Velu et al., 2008), rice (Krishnan et al., 2001), and wheat (Ozturk et al., 2006). The average chemical cost per sample is roughly $0.75 (given that 10 g of DTZ costs about $25 and 1000 ml of pure methanol costs about $12). For instance, a lab technician gets US$ 225 per month in pay, the cost of staff per sample would increase by US$ 0.1 consequently an additional cost of US$ 0.85 per sample. This protocol is basic and quick and affordable. This is a qualitative method that were not able to clearly distinguish the lines with average Zn content from those having high Zn content but good to use for discarding low groups in the preliminary selection.

Atomic Absorption Spectrophotometry (AAS): It is based on an atom's capacity to absorb light at extremely particular wavelengths. The precise quantitative determination of individual elements is made possible by the employment of clever light sources and reflective wavelength selection. The amount of light at the resonance wavelength that is absorbed as it travels through an atom cloud constitutes the element concentration of interest. The amount of light absorbed rises predictably with the number of atoms in the light path. One can quantify the amount of analyte element present in the sample by measuring the amount of light that is absorbed (Beaty & Kerber, 1993; Handson & Shelley, 1993). AAS requires sample digestions (i.e. solid samples to be brought into solution before measurements) and does multiple nutrients including Zn at different wavelengths. The price per sample varies US$4 -US$8 per mineral. AAS's daily capacity is nearly 100 digested samples. A well-established lab and a highly competent technician with costly and hazardous consumables are needed to accomplish this method.

Inductively coupled plasma optical emission spectroscopy (ICP-OES): This method estimates emission-based spectrometric technique wherein the excited electrons when returning to the ground state, emit energy at a specific wavelength (e.g. ~214 nm used for Zn prediction). The target element emits energy at a definite wavelength that is specific to its chemical makeup. ICP-OES typically chooses a single wavelength for a given element, even though each element emits energy at a variety of wavelengths. The amount (concentration) of the Zn in the sample under study determines the intensity of the energy released at the selected 213.857 nm wavelength (Skoog et al., 1995). To determine the quantity of Zn in samples (tissue or grain), this technique is particularly sensitive for the identification and quantification of soil pollutants (Al, Cr, and Ti) that come from sample preparations and helps to understand sample quality. The average subsidized cost of ICP-OES is US$18+ per sample, which can test 80–100 samples per day. Like AAS, this also necessitates a reputable laboratory, a highly skilled technician, hazardous consumables, and time-consuming sample preparation.

NIRS: The NIRS approach depends on the variation in chemical and physical components among various samples (Ludwig & Khanna, 2001). The stretching and vibrations of several bonds, including C–H, N–H, S–H, C–C, C=C, C–N, and O–H in the NIR spectral range, are responsible for the variation between samples in the amount and combination of absorption energy (700–2500 nm) (Foley et al., 1998). Because these NIR-active linkages are often found in organic substances like proteins and carbohydrates, NIRS is well suited for the study of these chemicals (Galvez-Sola et al., 2015; Ludwig & Khanna, 2001). Different bonds interact with NIR radiation in different ways, creating spectra with distinctive combinations of peaks and troughs that are exclusive to a sample (Richardson et al., 2004). Even though NIRS has a lot of potential for nutritional analysis, it has primarily been applied to the study of leaf samples and has hardly been applied to grain samples. Previous studies reported that NIRS generally overestimated the Zn content by up to 55% as compared to the ICP-OES method (Rai et al., 2012; Govindaraj et al. 2016a, b). The correlation between the ICP-OES and NIRS was weak (r = 0.24–0.58) for Zn content. Hence, not recommended for breeding sample screening and decision-making.

X–ray fluorescence (XRF): XRF runs on the principle that each element emits a secondary X-ray (fluorescent) with characteristic energy when exposed to X-rays of the proper higher energy, which in turn the intensity and energy level of the emitted X-rays are characteristics of the elemental composition in the sample. In ED-XRF, detectors are used which discriminate X-rays based on energy, enabling

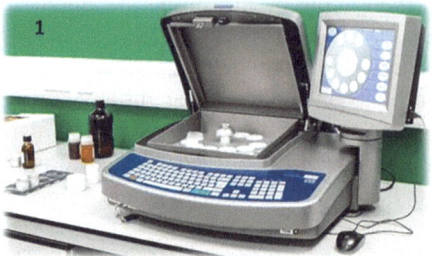
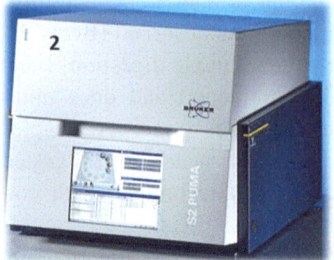

1. ED_XRF used for Rice and Pearl millet Zn analysis (Oxford X-Supreme 8000)

2. ED_XRF used for Wheat Zn analysis (Bruker)

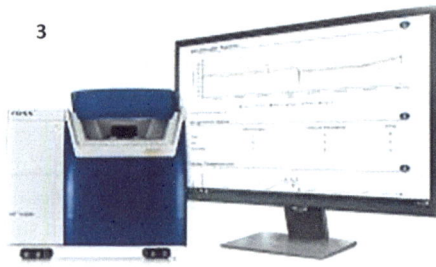

3. NIRS Near Infrared Spectrometer

4. 1,5-diphenyl-thio-carbazone staining method

Fig. 2 High throughput Zn phenotyping instruments. (1) Oxford X- Supreme 8000 ED-XRF bench top model; (2) Bruker ED-XRF bench top version; (3) NIRS- Near Infrared Spectrometer; (4) 1,5-diphenyl-thio-carbazone staining method

the simultaneous detection of multiple elements and it performs in a single run as ICP-OES does (Paltridge et al., 2012a, b). Setting up this small machine requires little space; its cost is less than US$ 70,000. The XRF distinguishes from AAS and ICP-OES in that it is a non-destructive technique for grain/seed Zn analysis with the ability to analyse a greater number of samples per day, on average 250–300 samples per day. The approach requires little to no sample preparation (Fig. 2). The average Zn estimate cost per sample is $3–4 USD compared to ICP-OES method which is too costly (about US$18/sample), and analyses 50% of XRF capacity (Rai et al., 2012). XRF method scarcely overestimated Zn content by 3–5%, The correlations between the ICP-OES and XRF values were highly positive and significant for Zn content ($r \geq 0.95$; $P < 0.01$). Based on these merits and rapid screening, globally more than 25 XRF facilities (Fig. 3) were established (mostly donated to NARS & CG centres) by the HarvestPlus program of the CGIAR, which enabled estimating Zn in more than 500,000 samples of cereals and legumes. For instance, pearl millet biofortification program handles 12,000–15,000 grain samples every year (Govindaraj et al. 2016a, b). Measured Zn expressed in ppm (mg/kg). To inspect XRF precision, each day (as a routine process) validation of check entry (known entry for Zn density) sample analysed and confirmed Zn values (with ± SE). XRF results are highly reproducible across labs and demonstrated by glass standards calibrations.

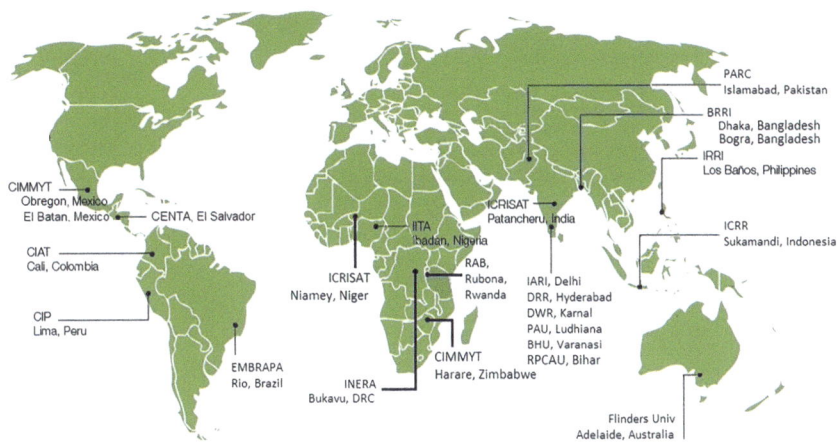

Fig. 3 XRF facilities established around the global south by HarvestPlus funded projects

Comparison of Different Methods for Zn Estimation

The comparisons between the various analytical techniques, its precision cost-effectiveness and their capacity are described in Table 1. XRF has shown equally competitive to ICP-OES performance by producing data more quickly and with steadier with a non-destructive way. The XRF Zn results among crops like pearl millet (Paltridge et al., 2012a, b; Rai et al., 2012; Stangoulis, 2010), rice (Paltridge et al., 2012a, b; Stangoulis, 2010) and wheat (Stangoulis, 2010 are in good agreement with ICP-OES. However, XRF needs to be calibrated initially using the established micronutrient criteria against the available crop variation spectrum. The response of the XRF to pure elements in a specific sample could be modelled by calibration using a wide range of sample types and sophisticated mathematical techniques (Perring & Andrey, 2003; Perring et al., 2005; Perring & Blanc, 2007, 2008; Rousseau et al., 1996). These XRF calibration techniques were created by Paltridge et al. (2012a, b) and are now frequently used for determining the micronutrient levels in wheat, rice, maize, pearl millet and beans samples. Although ICP-OES values were more accurate and precise, however, XRF method with a wide range of genetic materials applications as non-destructive makes an efficient method in large no. of breeding samples and gene bank accessions at reduced cost. The availability of XRF within the breeding program or the accessibility of such lab accelerates Zn-rich parents and progenies selection alongside agronomic traits assessment each generation of its advancement. Comparative analysis of ICP-OES and XRF for ZN in pearl millet accession showed significant positive correlations suggesting XRF-based selection significantly enhanced the Zn levels in the advanced progenies. Several progenies selection exercised in pearl millet for Zn using the XRF as a rapid method of selection (Fig. 4). The variability captured in ICP (20–70 ppm) and XRF (22–72 ppm) are similar range and confirm the higher precision for reproducibility.

Table 1 Comparison between different zinc (Zn) analytical methods and capacity

Subject	Staining	ICP-OES	AAS	XRF	NIRS
Method	Destructive	Destructive	Destructive	Non-destructive	Non-destructive
Sample preparation	Involves standard sample preparation	Involves typical sample preparation, which takes additional time	Involves typical sample preparation, which takes additional time	Minimal or no sample preparation	Minimal or no sample preparation
Calibration requirement	No calibration is required	No calibration is required	No calibration is required	Calibration is required	Calibration is required
Use of chemicals	Needs chemical compound 1,5-diphenyl-thio-carbazone	Involves acid digestion	Involves digestion	No chemicals involved	No chemicals involved
Quantification of Zn	Qualitative scored as low, medium, and high	Zn is quantified in ppm or mg kg^{-1}	Zn is quantified in ppm or mg kg^{-1}	Zn is quantified in ppm or mg kg^{-1}	Zn is quantified in ppm or mg kg^{-1}
High throughput	Can readily be used to remove low Zn lines based on the intensity of the stain	100 samples per day	100 samples per day	250–300 samples per day (crop dependent)	150 samples per day
Ability to detect and quantify typical soil contaminants	Cannot determine the contamination	Detect contamination by quantifying Al, Cr, Ti	Cannot determine the contamination	Can detect contamination by quantifying Al. (recent update)	Cannot determine the contamination
Reproducibility	Low	High	High	High	Moderate
Operation skill	Training	High skill set and training	High skill set and training	Simple	Simple

Improving XRF Method Accuracy

The recent development in the XRF lab on improving the data quality, reveals that the quality of the samples (free from contamination) determines the accuracy of the Zn data in any crop. A study consisting of several test samples from pearl miller rice and wheat indicates the samples that underwent distilled water wash and no wash had highly correlated results for Zn (Govindaraj, unpublished). The soil contamination did not show any significant effect on grain Zn density estimation, thus water washing of samples may not be required in Zn estimation while it is recommended

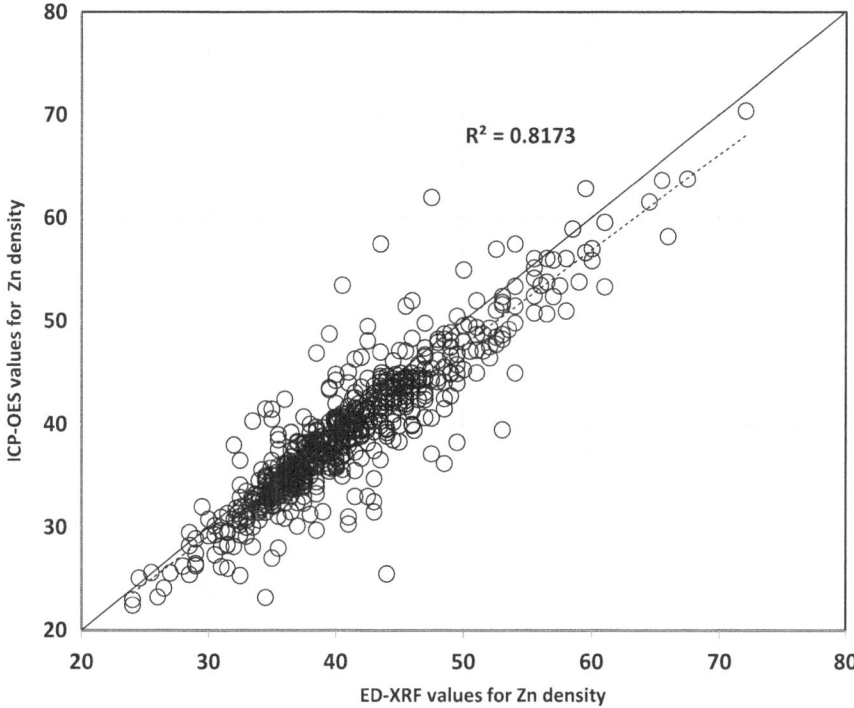

Fig. 4 Linear correlation between ICP-OES and XRF Zn data among 500 pearl millet accessions

for Fe estimation. Non-matrix matched glass disk standards-based calibration improves the XRF mineral analysis in rice, wheat, maize, pearl millet, sorghum, beans and lentils (Guild & Stangoulis, 2016). This helps in improving the sample prediction accuracy and accounts for small variations between the instruments or labs for better reproducibility using these calibrations. For instance, Wheat validation samples were tested using the glass standards calibrations across five XRF instruments in India. Results showed differences between the XRF labs depicted in Fig. 5, inferring an improvement in the Zn estimation accuracy (as per correlation coefficient) and improvement in reproducibility between instruments or labs when comparing the results with the glass calibration method. Difference between XRF and ICP-OES results is likely expected due to the inherent variability caused by sampling whole grain and the non-destructive nature of the analysis. Therefore, sample reproducibility can be improved with glass calibrations, longer scan times, replicate analyses and ground samples XRF analysis. However, high throughput screening is the aim of this method and improved calibrations with single replicas is adequate for this application (Paltridge et al., 2012a, b; Guild & Stangoulis, 2016). Over the years, several thousand samples were screened for Zn in rice and wheat as primary target crops (Fig. 6a, b). Genetic variation for Zn is 10–68 ppm in wheat samples and 7.5–29 ppm in rice samples. Wheat results indicate better progress

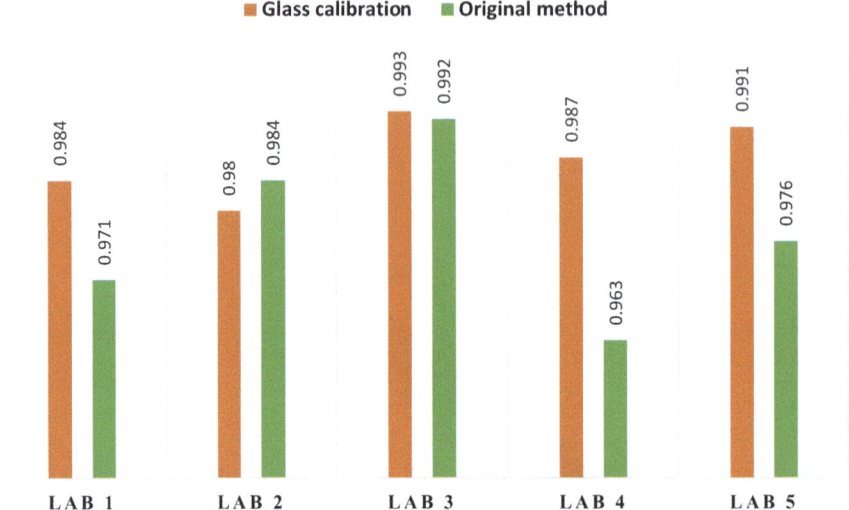

Fig. 5 Correlation between ICP-OES and XRF results for Zn generated using glass calibration and grain calibration in wheat validation samples at different labs in India

toward global biofortification breeding target (≥37 ppm Zn) than rice targets (≥28 ppm Zn). The nature of samples used, and grain consumed may be a factor-in for rice as polished.

XRF Constraints and Opportunities

Although XRF is a novel analytical technique that supports mineral estimation in various crops, it has some limitations to improve efficacy in the coming years. Screening larger grain samples (cowpea, beans, and maize) than wheat/rice/millet, requires at least semi-destructive (grits to flour) to improve the precision of estimates. This is because increasing seed diameter will result in fewer seeds being exposed to x-rays which will also make the sample being analysed less representative. Additionally, the larger grains' spherical geometry would reduce packing density and increase air cross-section in the samples of these larger seeds that were studied when compared to smaller grains with large diameters (Donev et al., 2004; Weitz, 2004; Li et al., 2010). Use of grits and flour samples necessitates changing cups between each scan besides additional labour costs for grind samples but still cost-effective than existing methods. On the other side, the sample must be completely free of contamination, as many XRF labs find it difficult to map those samples or misinterpret them as high values. Such a case is applicable for iron (Fe) in grain but not for Zn estimates. Researchers continuously improve their understanding of XRF usage and sampling. A millet study reported high Zn lines selection from the 11 trials (>1000samples), it was possible to conclude that selecting high

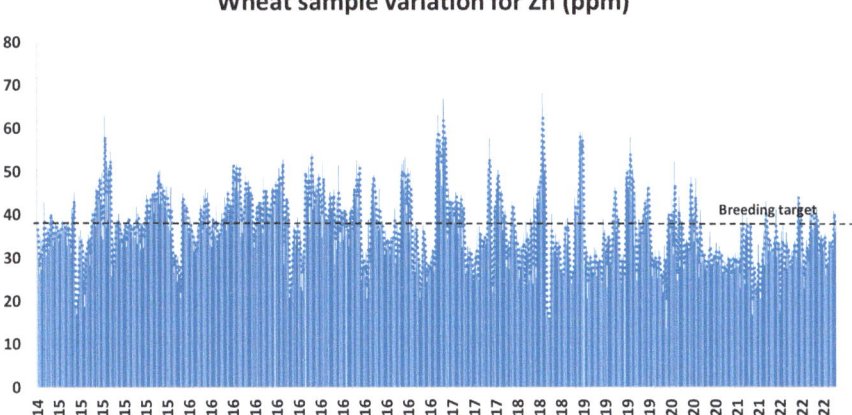

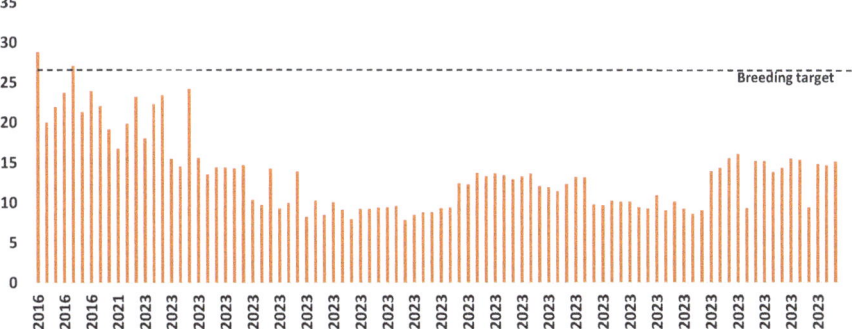

Fig. 6 (**a**) Various wheat samples analyzed (2014–2022) using XRF for Zn content and acquired genetic variation (the dotted line represents Zn global breeding target, 37 ppm). (**b**) Various rice samples analyzed (2016–2023) using XRF for Zn content and acquired genetic variation (the dotted line represents Zn global breeding target, 28 ppm)

Zn lines above the trial mean would typically pick roughly 50% of the genotypes for the subsequent testing stage (Govindaraj et al., 2016a, b). Further, such XRF-based selection for grain Zn, indicates high levels of consistency of the ranking of entries across the test environments for Zn. Therefore, the XRF method is strongly recommended for Zn analysis (in addition to Fe) as a quick, reliable, and high throughput method for very small sample sizes non-destructive way (4–6 g). The commercial cost of XRF is much cheaper (5–6 USD) than other methods. This cost is likely to be reduced up to 50% in the coming years if the total capacity (~40,000 samples) of the lab is effectively used annually. HarvestPlus offers cost-effective capacity and screening; interested partners advised to explore the online service portal (https://harvestplus.solutions/my-account/).

Conclusion

XRF has emerged as the non-destructive, accurate, and proven high-throughput technology for crop biofortification programs, despite the fact that there are other ways to phenotype Zn micronutrients in cereals. This innovative technique is essential to the global and regional crop improvement initiatives because it speeds up the process of screening breeding populations and advancing the stage gate of Zn biofortified product development, testing, and release for widespread cultivation. Strong positive correlation between ICP and XRF, it suggests that the XRF readings accurately reflect sample fluctuations. The introduction of glass standards, which are the gold standard for XRF reproducibility both within and outside of the laboratory, for data accuracy and breeding program decision-making. A distinct advantage is proving that it is feasible to employ XRF for early-stage screening of breeding material (limited seeds) in order to reject seeds with low Zn content and use the same/remaining seeds for planting. Numerous countries and crops have benefited from XRF's analysis of millions of samples over the decade, which has led to the production of over 300 biofortified cultivars. With spectrum analysis graphs, XRF can also scan many minerals at once, which is an intriguing prospect for the future. Therefore, in large-scale screening and fast-track breeding operations involving public and commercial companies, XRF results in higher throughput and significant cost savings method for mainstreaming nutrition CGIAR and NARES programs.

References

Beaty, R. D., & Kerber, J. D. (1993). *Concepts, instrumentation and techniques in atomic absorption spectrometry*. Perkin-Elmer.

Bouis, H. E. (2003). Micronutrient fortification of plants through plant breeding: Can it improve nutrition in man at low cost. *Proceedings of the Nutrition Society, 62*, 403–411.

Dary, O., & Mora, J. O. (2002). Food fortification to reduce vitamin A deficiency: International vitamin A consultative group recommendations. *The Journal of Nutrition, 132*(9), 2927–2933. https://doi.org/10.1093/jn/132.9.2927S

De Romaña, Bown, D. K., & Giunard, J. X. (2002). Sensory trial to assess the acceptability of zinc fortificants added to iron-fortified wheat products. *Journal of Food Science, 67*(1), 461–465. https://doi.org/10.1111/j.1365-2621.2002.tb11429.x

Donev, A., Cisse, I., Sachs, D., et al. (2004). Improving the density of jammed disordered packings using ellipsoids. *Science, 303*, 990–993. https://doi.org/10.1126/science.1093010

Foley, W. J., Mcilwee, A., Lawler, I., Aragones, L., Woolnough, A. P., & Berding, N. (1998). Ecological applications of near infrared reflectance spectroscopy—a tool for rapid, cost-effective prediction of the composition of plant and animal tissues and aspects of animal performance. *Oecologia, 116*, 293–305.

Galvez-Sola, L., Garcı'a-Sa'nchez, F., Perez-Perez, J. G., Gimeno, V., Navarro, J. M., Moral, R., Martı'nez-Nicola's, J. J., & Nieves, M. (2015). Rapid estimation of nutritional elements on citrus leaves by near infrared reflectance spectroscopy. *Frontiers in Plant Science, 6*, 571.

Govindaraj, M., Rai, K. N., Pfeiffer, W. H., Kanatti, A., & Shivade, H. (2016a). Energy-dispersive x-ray fluorescence spectrometry for cost-effective and rapid screening of pearl millet germplasm and breeding lines for grain iron and zinc density. *Communications in Soil Science and Plant Analysis, 47*, 2126–2134. https://doi.org/10.1080/00103624.2016.1228938

Govindaraj, M., Rai, K. N., & Shanmugasundaram, P. (2016b). Intra-population genetic variance for grain iron and zinc contents and agronomic traits in pearl millet. *Crop Journal, 4*(1), 48–54.

Guild, G. E., & Stangoulis, J. C. R. (2016). Non-matrix matched glass disk calibration standards improve XRF micronutrient analysis of wheat grain across five laboratories in India. *Frontiers in Plant Science, 7*, 784. https://doi.org/10.3389/fpls.2016.0078

Handson, P. D., & Shelley, B. C. (1993). A review of plant analysis in Australia. *Australian Journal of Experimental Agriculture, 33*(8), 1029–1038. https://doi.org/10.1071/EA9931029

HarvestPlus. (2008). HarvestPlus technical monographs. In J. Stangoulis & C. Sison (Eds.), *Crop sampling protocols for micronutrient analysis*. International Food Policy Research Institute.

Krishnan, S., Ebenezer, G. A. I., & Dayanandan, P. (2001). Histochemical localization of storage components in caryopsis of rice (*Oryza sativa* L.). *Current Science, 80*, 567–571.

Li, S., Zhao, J., Lu, P., & Xie, Y. (2010). Maximum packing densities of basic 3D objects. *Chinese Science Bulletin, 55*, 114–119. https://doi.org/10.1007/s11434-009-0650-0

Ludwig, B., & Khanna, P. (2001). Use of near infrared spectroscopy to determine inorganic and organic carbon fractions in soil and litter. In *Assessment methods for soil carbon* (pp. 361–370). Lewis Publishers.

Ozturk, L., Yazici, M. A., Yucel, C., Torun, A., Cekic, C., Bagci, A., Ozkan, H., Braun, H. J., Sayers, Z., & Cakmak, I. (2006). Concentration and localization of zinc during seed development and grain germination in wheat. *Physiologia Plantarum, 128*(1), 144–152. https://doi.org/10.1111/j.1399-3054.2006.00737.x

Paltridge, N. G., Milham, P. J., Ortiz-Monasterio, J. I., et al. (2012a). Energy-dispersive X-ray fluorescence spectrometry as a tool for zinc, iron and selenium analysis in whole grain wheat. *Plant and Soil, 361*, 261–269. https://doi.org/10.1007/s11104-012-1423-0

Paltridge, N. G., Palmer, L. J., Milham, P. J., et al. (2012b). Energy-dispersive X-ray fluorescence analysis of zinc and iron concentration in rice and pearl millet grain. *Plant and Soil, 361*, 251–260. https://doi.org/10.1007/s11104-011-1104-4

Perring, L., & Andrey, D. (2003). ED-XRF as a tool for rapid minerals control in milk-based products. *Journal of Agricultural and Food Chemistry, 51*(15), 4207–4212. https://doi.org/10.1021/jf034158p

Perring, L., & Blanc, J. (2007). EDXRF determination of iron during infant cereals production and its fitness for purpose. *International Journal of Food Science, 42*(5), 551–555. https://doi.org/10.1111/j.1365-2621.2006.01265.x

Perring, L., & Blanc, J. (2008). Validation of quick measurement of mineral nutrients in milk powders: Comparison of energy dispersive X-ray fluorescence with inductively coupled plasma-optical emission spectroscopy and potentiometry reference methods. *Sensing and Instrumentation for Food Quality and Safety, 2*(4), 254–261. https://doi.org/10.1007/s11694-008-9056-y

Perring, L., Andrey, D., Basic-Dvorzak, M., & Hammer, D. (2005). Rapid quantification of iron, copper and zinc in food pre- mixes using energy dispersive X-ray fluorescence. *Journal of Food Composition and Analysis, 18*(7), 655–663. https://doi.org/10.1016/j.jfca.2004.06.011

Prasad, A. S. (2013). Discovery of human zinc deficiency: Its impact on human health and disease. *Advances in Nutrition, 4*(2), 176–190. https://doi.org/10.3945/an.112.003210

Rai, K. N., Govindaraj, M., & Rao, A. S. (2012). Genetic enhancement of grain iron and zinc content in pearl millet. *Quality Assurance and Safety of Crops & Foods, 4*(3), 119–125. https://doi.org/10.1111/j.1757-837X.2012.00135.x

Rai, K. N., Yadav, O. P., Rajpurohit, B. S., Patil, H. T., Govindaraj, M., Khairwal, I. S., Rao, A. S., Shivade, H., Pawar, V. Y., & Kulkarni, M. P. (2013). Breeding pearl millet cultivars for high iron density with zinc density as an associated trait. *Journal of SAT Agricultural Research, 11*, 1–7.

Richardson, A. D., Reeves Iii, J. B., & Gregoire, T. G. (2004). Multivariate analyses of visible/near infrared (VIS/NIR) absorbance spectra reveal underlying spectral differences among dried, ground conifer needle samples from different growth environments. *The New Phytologist, 161*, 291–301.

Rousseau, R. M., Willis, J. P., & Duncan, A. R. (1996). Practical XRF calibration procedures for major and trace elements. *X-Ray Spectrometry, 25*(4), 179–189. https://doi.org/10.1002/(SICI)1097-4539(199607)25:4<179::AID-XRS162>3.0.CO;2-Y

Skoog, D. A., West, D. A., & Holler, F. J. (1995). *Química Analítica* (6th ed.). Mc. Graw Hil.

Stangoulis, J. C. R. (2010). *Technical aspects of zinc and iron analysis in biofortification of the staple food crops, wheat and rice*. 19th World congress of soil science, soil solutions for a changing world, 1–6 August 2010, Brisbane.

Stangoulis, J., & Sison, C. (2008). *Crop sampling protocols for micronutrient analysis*. HarvestPlus Technical Monograph Series 7, ISBN 978-0-9818176-0-6.

Velu, G., Bhattacharjee, R., Rai, K. N., Sahrawat, K. L., & Longvah, T. (2008). A simple and rapid screening method for grain zinc content in pearl millet. *Journal of Agricultural Research, 6*, 1–4.

Weitz, D. A. (2004). Packing in the spheres. *Science, 303*, 968–969. https://doi.org/10.1126/science.1094581

Zuo, Y., & Zhang, F. (2009). Iron and zinc biofortification strategies in dicot plants by intercropping with gramineous species. *Agronomy for Sustainable Development, 29*(1), 63–71. https://doi.org/10.1051/agro:2008055

Open Access This chapter is licensed under the terms of the Creative Commons Attribution 4.0 International License (http://creativecommons.org/licenses/by/4.0/), which permits use, sharing, adaptation, distribution and reproduction in any medium or format, as long as you give appropriate credit to the original author(s) and the source, provide a link to the Creative Commons license and indicate if changes were made.

The images or other third party material in this chapter are included in the chapter's Creative Commons license, unless indicated otherwise in a credit line to the material. If material is not included in the chapter's Creative Commons license and your intended use is not permitted by statutory regulation or exceeds the permitted use, you will need to obtain permission directly from the copyright holder.

Breeding and Deployment of High Zn Wheat in South Asia

Velu Govindan, Arun Kumar Joshi, Pradeep Bhati, and Karthikeyan Thiyagarajan

Introduction

Wheat is an important dietary source of a range of components which is essential for human nutrition and health. These include mineral micronutrients (notably iron, zinc and selenium but also calcium and magnesium), vitamins (notably B vitamins), dietary fibre (DF) and phytochemicals with putative health benefits (notably phenolic acids). Although the extent to which wheat contributes to the total intakes of these components varies between countries (in relation to wheat intakes and other dietary sources), there is no doubt that deficiencies of iron, zinc, B vitamins and DF are global problems.

More than two billion people around the world suffer from micronutrient deficiencies, which are collectively known as "hidden hunger." The name refers to the fact that many of the symptoms of micronutrient deficiencies are not easy to see but can have a detrimental lifelong impact, such as impairment of mental capacities, lower resistance to diseases. Around 17% of the global population (roughly 1.3 billion people) are at risk of inadequate zinc intake (WHO, 2021). The prevalence of zinc deficiency is estimated to exceed 25% in sub-Saharan Africa and 29% in South Asia. Since zinc deficiency is a cause of stunting (low height for age), stunting is commonly used as a proxy to estimate the risk of zinc deficiency in a population (Global Nutrition Report, 2018). Approximately 23% of all preschool-age children are stunted. The diets of low-income consumers in developing countries usually

V. Govindan (✉)
International Maize and Wheat Improvement Center (CIMMYT), Texcoco,
Estado de México, Mexico
e-mail: velu@cgiar.org

A. K. Joshi
Borlaug Institute for South Asia (BISA), New Delhi, India

P. Bhati · K. Thiyagarajan
Borlaug Institute for South Asia (BISA), Ludhiana, India

consist of larger amounts of staple foods (such as wheat, maize and rice) and fewer micronutrient-rich foods such as fruits, vegetables, and animal products. The double burden of disease is where undernutrition exists along with diet-related noncommunicable diseases (NCDs) such as overweight and obesity. This double burden also exists in low- and middle-income countries, and zinc may play a role in addressing it. A meta-analysis showed that on the relationship between zinc supplementation and risk factors for two common NCDs: type 2 diabetes and cardiovascular disease. Research indicates that low-dose, long-duration intake of zinc through supplements reduced risk factors for these NCDs—raising the possibility that consumption of zinc biofortified foods might have the same benefit.

Zinc is involved in more bodily functions than any other mineral. Zinc is essential to more than 200 enzyme systems, normal growth and development, the maintenance of body tissues, reproductive health, vision, and the immune system. Zinc is vital for survival, meaning its deficiency has serious consequences for health, particularly during childhood when zinc requirements are higher. In addition to stunting, zinc deficiency can increase the risk of common childhood infections, including diarrhea, pneumonia, and malaria. Most diets (especially in low-income countries) do not contain enough zinc, making zinc deficiency one of the biggest causes of hidden hunger globally. For example, inadequate zinc intake in India is partly responsible for a 35% rate of stunting among children younger than 5 years, which also leads to frequent infections and inflammation in this age group.

Physiological Functions of Zinc and Iron and Their Status

Zinc can neither be synthesised in the human body nor stored in substantial amounts and hence requires a regular intake through diet (Maxfield et al., 2022). The recommended dietary allowance (RDA) of Zn per day varies from 3 mg in children to 8 and 11 mg in adult females and males respectively. However, for pregnant and lactating women the RDA of zinc is around 12 and 13 mg per day respectively, making these the most susceptible group in the case of poor zinc intake (Institute of Medicine, 2001). More than 300 enzymes and 2000 transcriptional factors are Zn-dependent and have a vital role in the replication and repair of DNA, immunity, wound healing, fertility, foetal development, metabolism of biomolecules and mental health (Chasapis et al., 2020). Zinc deficiency is ranked fifth among the leading risk factors for illness and disease in developing countries with high mortality (International Zinc Nutrition Consultative Group (IZiNCG) et al., 2004). It is estimated that nearly 17.3% of the global population is affected by Zn deficiency. Zinc deficiency shows strong socio-economic disparity, ranging from 7.5% in high-income regions to 30% in South Asia, with more than 20% of prevalence in LMIC's (Wessells & Brown, 2012; Gupta et al., 2020). Zinc was responsible for 14.4%, 10.4% and 6.7% of child deaths linked to diarrhoea, malaria and pneumonia respectively, and accounted for nearly 0.45 million children's deaths in Africa, Asia, and Latin America in the year 2004 (Fischer Walker et al., 2009).

Measures in Food Systems to Tackle Zinc Deficiency

Efforts are being made to address zinc and iron "hidden hunger" by various means, such as the diversification of diets, supplementation, biofortification, and the fortification of commercial foods. However, supplementation and fortified foods have not been found to be of much use in tackling the problem of zinc and iron deficiency because of their increased cost and lack of acceptance by consumers. In addition, supplementation and fortification measures require recurrent investments and thus remain unsustainable (Lockyer et al., 2018; Pfeiffer & McClafferty, 2007).

Biofortification refers to improving the status of health-beneficial micronutrients in food crops; this can be achieved through two approaches, namely agronomic and genetic (conventional breeding methods and gene modification) (Velu et al., 2014). Agronomic biofortification uses fertilizer management strategies such as soil and foliar applications of micronutrient fertilizers to improve the micronutrient status of the food. There is evidence of increased micronutrient concentration with agronomic biofortification: for instance, the foliar application of $ZnSO_4$ leads to an increase of about 60% in grain zinc concentration in wheat (Zhang et al. 2012). However, many farmers in LMIC's cannot afford the cost of micronutrient fertilizers. In addition, agronomic biofortification may not be environmentally acceptable or sustainable and may serve only as a short-term solution, as it requires the constant input of resources for every crop season.

The genetic biofortification approach uses plant breeding techniques for increasing micronutrient density, substances that promote nutrient absorption and reduce the levels of anti-nutrients in staple food crops (Bouis, 2003). It encompasses both traditional breeding and gene modification methods. While gene-editing techniques may result in high micronutrient-dense cultivars, consumer and regulatory acceptance is not universal. Contrarily, biofortification achieved through the traditional breeding approach has been shown to improve the levels of health-beneficial micronutrients, besides having wide acceptability. The major costs incurred in the conventional plant breeding approach are related to the research required in breeding for biofortified varieties and implementation in the initial stages (Nestel et al., 2006). In the next stages, after a few promising micronutrient-dense cultivars are developed, they can be incorporated into germplasm and breeding pipelines specific to production zones, an approach popularly referred to as "mainstreaming". Once the mainstream breeding of biofortified cultivars is accomplished, minimal resources are required for maintenance breeding of the biofortification traits (Virk et al., 2021). Therefore, biofortification through conventional breeding offers a sustainable, efficient, long-term, and environmentally friendly solution to the development of micronutrient-dense crop varieties by exploiting natural genetic diversity (Pfeiffer & McClafferty, 2007). In addition, it can be synergistically coupled with other approaches such as agronomic biofortification, supplementation and fortification to tackle zinc and iron deficiency in LMIC's (Velu et al., 2014). For simplicity, biofortification through conventional plant breeding will henceforth be referred to as biofortification.

Thus, plant breeding offers a promising strategy to contribute to the improvement of the nutritional status of staple food crops across the globe. To develop micronutrient-dense varieties, the National Agricultural Research Systems (NARS) are collaborating with international organizations such as partner institutes of the Consultative Group on International Agricultural Research (CGIAR). HarvestPlus (HarvestPlus) is a part of the CGIAR research program, leading biofortification research into various staple food crops such as common bean, cassava, maize, rice, sweet potato, and wheat, with the interdisciplinary collaboration of different NARS, CGIAR, and their extension centres, and with primary focus on LMIC's. With the remarkable research efforts in biofortification, several biofortified varieties of staple crops have been commercially released and/or are in the testing stages across the world (Biofortified Crops Around the World_0.pdf (harvestplus.org)).

The Role of Wheat in Nutrition

Wheat

Wheat is one of the most important staple foods, with the highest cultivation area (225 million ha) and the second-highest production quantity (770 million tonnes) in the world (FAO et al., 2021). Wheat accounts for nearly 8% of global crop production and ranks second after maize. Asia leads in the production of wheat, accounting for 44% of global production, followed by Europe and America. According to OECD (2021), global wheat production area and yields are expected to rise by 3% and 10% respectively by 2030. China, the European Union, India, Russia, and the United States in their respective order are expected to be the top producers of wheat and together by the year 2030 account for nearly 63% of global wheat production. It is anticipated that the top consumers, on the other hand, will be China, India, the European Union, Russia, and the United States in their respective order, and together will account for 52% of wheat consumption globally. According to the International Wheat Genome Sequencing Consortium (IWGSC) (2014), common wheat, often known as bread wheat, accounts for around 95% of all wheat production worldwide. Wheat is consumed by nearly 30% (2.5 billion) of the population globally (FAO et al., 2021), and accounts for nearly 18% (530 kcal) of human calorie intake and 19% of protein intake for the triennium ending 2017 (Erenstein et al., 2022).

Wheat Grain in Relation to Zinc and Iron

The dry weight of the endosperm consists of 70–80% starch, 10–15% protein, 4–5% dietary fibre and 2–3% lipids. The starchy endosperm also contains other elements such as phytochemicals, minerals and vitamins. The aleurone layer and embryo are rich in fibre, protein, minerals, B vitamins, and phytochemicals (Gooding & Shewry,

2022; Hazard et al., 2020) However, the minerals from the aleurone layer and embryo are bound to phytate, which reduces the bioavailability of minerals in the digestive tract (Bechtel et al., 2009; Roohani et al., 2013). The fibre-rich outer layers protect the grain and are mostly lost in the milling process when the extraction rate of bran (refinement rate) is higher. While Fe is mostly concentrated in the aleurone layer and the scutellum region of the embryo, Zn on the other hand is concentrated in the embryonic axis of the embryo and in the epithelium of the scutellum (Wan et al., 2022). Bouis and Welch (2010) found that the bioavailability of Zn and Fe in wholegrain wheat to be around 25% and 10% of the total Zn and Fe respectively. The phytates are concentrated mostly in the outer layers, and this explains why the Zn, which is concentrated in the embryonic axis and linked to other enzymes and proteins instead of phytates, has better bioavailability (Broadley et al., 2007). According to Hussain et al. (2013), better concentrations of Zn and Fe were retained in white flour in the case of biofortified zinc compared to the non-biofortified variety (Balk et al. 2019), and this property increases with an increase in extraction rate. Also, Zn and Fe bioavailability is substantially increased by biofortification, irrespective of the extraction level (Rosado et al., 2009).

Breeding for Zinc Biofortification in Wheat

Breeding wheat with higher Zn and Fe concentrations offers a promising, practical, and potentially low-cost sustainable solution to prevent or curb human Zn and Fe deficiencies in LMIC's (Singh et al., 2017). The International Maize and Wheat Improvement Centre (CIMMYT), Mexico, in partnership with HarvestPlus and National Agricultural Research Systems (NARS), leads in the development of high-yielding and micronutrient-dense wheat varieties adapted to various agro-climatic environments. For instance, CIMMYT-bred high-Zn wheat varieties with up to 40 mg/kg of Zn adapted for South Asia would contribute up to 70% of the recommended daily zinc requirement.

Currently grown popular varieties of wheat contain about a 25-ppm baseline concentration of Zn and Fe in the whole wheat grain. The target for achieving high grain Zn and Fe is 37 ppm, which is 12 ppm more than the popular varieties contain (Bouis & Welch, 2010; Velu et al., 2022). The CIMMYT-derived first high-zinc wheat variety "Zinc-Shakti" released in India is a derivative of a synthetic hexaploid wheat crossed with an elite wheat parent, and Zinc-Shakti shows a stable performance with a 14 ppm more Zn advantage over local varieties and profitable yields, and also matures 2–3 weeks earlier than other varieties. For instance, high zinc wheat variety 'Akbar-19' grown over 40% area in Pakistan. Biofortified high zinc varieties cultivated in India and Pakistan have a 20–40% higher grain Zn with yields reasonably similar to the best local varieties (Velu et al., 2020).

The magnitude of Fe and Zn deficiency is particularly severe among children and pregnant and lactating women. Biofortified wheat with increased grain Zn and Fe has several potential advantages as a delivery vehicle of Zn and partially for Fe through wheat in South Asia and Ethiopia, and the Zn enriched wheat can provide up to 50% of daily recommended allowance for humans. Most of the wheat produced in the targeted regions is milled locally, and the use of whole grain wheat flour in food products allows retaining most of the zinc in the grain as these minerals are concentrated in the outer layer of the grain. The consumers in South Asia and Ethiopia prefer flatbreads, such as *chapatti, roti, nan,* and other wholegrain products including porridge.

Genetic Diversity and Targeted Breeding for Zinc

Large-scale screening of diverse genetic resources from CIMMYT germplasm bank and other sources have shown that there is a significant genetic variability for Zn and Fe content in some wheat genepools from primitive wheats, wild relatives and landraces, indicating that Zn content is amenable to rapid breeding progress. Landraces and wild relatives of common wheat such as *Triticum Spelta, T. dicoccon* and *T. turgidum* based synthetics that had the highest levels of Zn and Fe were used in targeted transfer using limited backcrossing into elite breeding lines.

Good progress has been made in the past decade in transferring alleles for high-zinc from these sources into elite breeding lines using large segregating populations grown in Toluca and Cd. Obregon environments in Mexico. Elite high Zn lines combining high Zn (and Fe), comparable yield potential, disease resistance, stress tolerance and quality were identified; high Zn varieties are released in India, Pakistan, Bangladesh, Nepal, Mexico, Ethiopia and Bolivia.

Targeted Breeding Approach to Enhance Grain Zn Content

The targeted breeding focused on simultaneous enhancement of Zn concentration and high yield has become the key objective after achieving success from the proof-of-concept approach. Each year about 400–500 simple crosses were made between elite high/moderate Zn lines, and between elite high Zn lines and best lines with normal Zn. Three-way crosses, or single back-crosses (BC1), were made with a high yielding parent. The BC1/F1Top and other segregating populations are shuttled between Obregon and Toluca field sites as described in breeding pipelines. In all generations, plants are selected for agronomic traits and disease resistance (all three rusts, Septoria tritici blight), 1–2 spikes from selected plants harvested as bulk, plump bold grains retained for advancing to next generation. Selected plants in the F4/F5 generations are harvested individually, selected for grain traits and grown as F5/F6 small plots for phenotyping. Lines retained for agronomic traits and disease resistance were harvested, selected for grain characteristics and grain Zn and Fe concentration determined using XRF machine. High Zn carrying F5/F6 lines are advanced to stage 1 replicated yield trials at Obregon in the Zn-homogenized fields, which has shown good prediction of grain Zn in South Asia and other TPEs. Lines that yield similar or better than the checks in stage 1 yield trials are analyzed for grain Zn and Fe, and selected lines analyzed for end-use processing quality. Line in stage 1 yield trials are also simultaneously phenotyped for resistance to Ug99 and yellow rust at Njoro, Kenya-off season, and the lines retained from Obregon trial again in the main-season. Seed multiplication of retained lines then conducted in El Batan while they are also phenotyped for rusts and other diseases. Competitive high Zn lines combined with key agronomic traits are distributed to NARS partners in South Asia and other TPEs. This led to identification and release of two dozens competitive high Zn varieties in TPEs in South Asia, Ethiopia and in Bolivia, these varieties are estimated to be grown by at least 10 million smallholder farmers.

A recent stage 1 yield trials data from Ciudad Obregon showed about 1% average yield gain achieved over the years while enhancing grain Zn concentration with +1 ppm annually, suggesting a high probability of combining high yield with high Zn concentration (Figs. 1 and 2). Although the mean yields of breeding lines derived from high Zn breeding pipeline and main breeding program were same, mean yield of 'selected lines' with high Zn values were 4–6% lower than the mean of 'selected lines' from main breeding program. Moreover, the lack of association between grain yield and grain Zn further supports their simultaneous genetic gain as realized in our current breeding scheme.

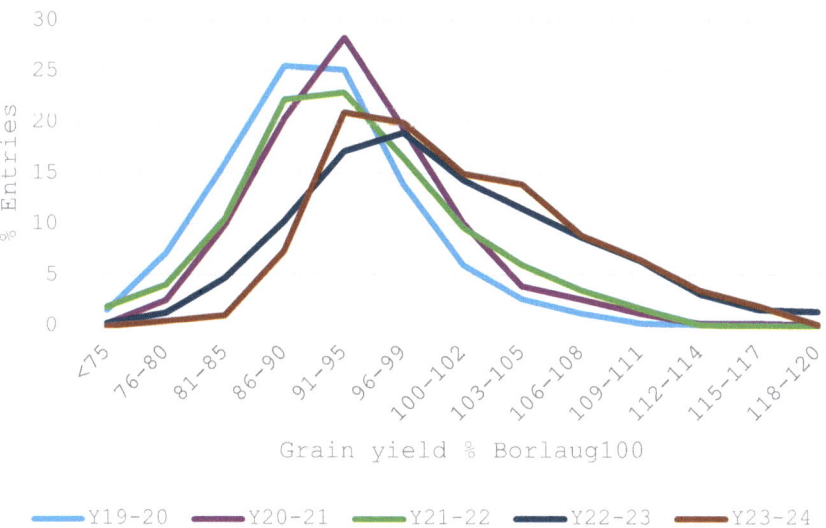

Fig. 1 Grain yield trends of wheat lines derived from three cohorts of Zn breeding pipeline evaluated in stage 1 replicated yield trials at Ciudad Obregon 2016–17 to 2023–24

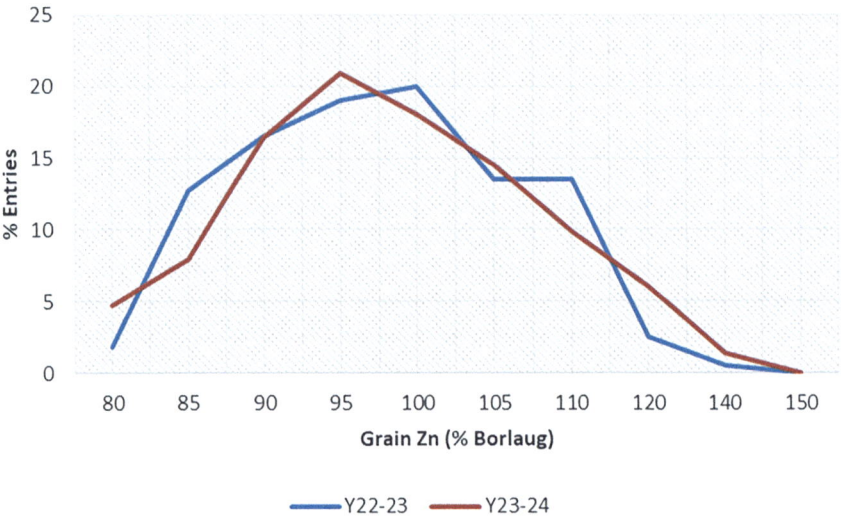

Fig. 2 Grain Zn concentration of wheat lines derived from two cohorts of Zn breeding pipeline evaluated in stage 1 replicated yield trials at Ciudad Obregon during 2022–23 and 2023–24

Zn Biofortification: Challenges and Opportunities

The major challenge over the next decades will be to maintain the rates of genetic gains for grain yield along with increased grain Zn concentration as well as to close the yield gap of 4–6% between non-biofortified vs biofortified lines. Therefore, to remain competitive, the performance of Zn-enhanced lines/varieties must be equal or superior to that of current non-biofortified elite lines/varieties, to ensure that smallholders will continue to adopt them. Since both yield and Zn content are invisible and quantitatively inherited traits except few intermediate effect QTL regions identified for grain Zn, increased breeding efforts and new approaches could combine them at high frequency in CIMMYT's elite germplasm, ensuring that Zn levels are steadily increased to the required levels across the CIMMYT breeding pipeline.

The addition of Zn as a core trait likely requires a significant acceleration in the breeding cycle, expanding population sizes, phenotyping for Zn, yield testing and expanded land use, phenotyping for biotic and abiotic stresses, genotyping, molecular-assisted selection and genomic selection. While continuing to increase agronomic performance, high Zn alleles will be added as a core trait and the Zn content will be increased in breeding lines annually along with the frequency of elite lines with high Zn with potential to be released by partners.

In addition, heterogeneity within experimental plots for available soil Zn remains a challenge. Our experimental fields at Ciudad Obregon was optimized using soil application of Zn fertilizer over the years. We are expanding this area to all fields where wheat-breeding materials are grown to allow reliable Zn phenotyping. Similar approaches will be followed in key sites in TPEs to optimize and improve the homogeneity for available soil Zn to allow reliable selection of lines with enhanced genetic potential to accumulate more Zn in grain.

New Breeding Schemes to Accelerate the Genetic Gain for Grin Zn

We have optimized two new breeding schemes for piloting, which have potential to accelerate simultaneous genetic gain for grain yield and zinc and permit Zn-mainstreaming in CIMMYT wheat germplasm.

Rapid Bulk Generation Advancement (RBGA) Scheme

An efficient 3-years rapid bulk generation advancement (RBGA) scheme have been optimized (Fig. 5). This has been achieved by targeted simple crosses made using candidate varieties and elite lines with high Zn sources, advanced under the new scheme, using the Toluca screen house facility (Fig. 3). We advance the F3 generation in the field in Toluca during the summer field season to conduct selection for agronomic traits, disease resistance and grain characteristics, harvest single spikes from selected plants with enough seed availability to grow short F4 plots. Resulting F3-derived F4 head-rows selected in Zn-enriched fields in Obregon will be followed by stage 1 (1 environment) and stage 2 grain yield performance testing, quality analysis, and disease phenotyping conducted in Mexico, South Asia, Ethiopia and Kenya to identify varietal candidates. The RBGA scheme combines field and Toluca screenhouse facilities that would allow completing a breeding cycle in 3 years by taking 2 years from the current shuttle breeding cycle of 5 years. The RBGA scheme allows to maintain large diversity within the segregating populations and to enhance efficiency, time and cost saving methods and manage large numbers of crosses for Zn mainstreaming and AGG projects.

One of the potential problems in the above selection scheme is the quality of seed obtained from 1 spike in Toluca to grow uniform F4 plots in Obregon due to high rainfall and Fusarium head blight infection in the field. In the past we have faced issues with it when managing large number of populations and made decision to harvest individual plants in Obregon instead of harvesting single spikes in Toluca, which adds an additional year but allowed better phenotyping for agronomic traits and discarding a larger number of lines as small plots in Toluca. In this case, generation advancement in RBGA scheme will take an additional year and will require 3 years instead of two as proposed in Fig. 3.

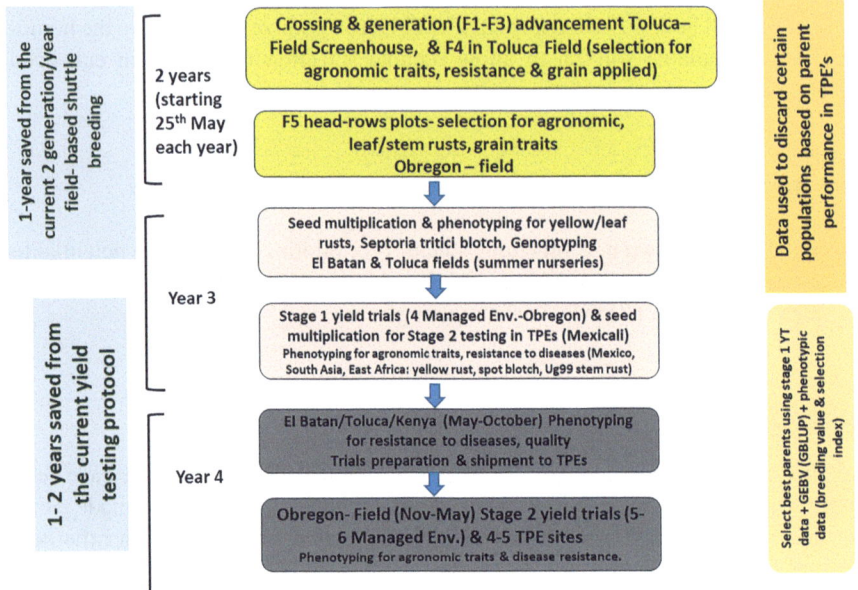

Fig. 3 New 3-years RBGA breeding scheme

Rapid-Cycle Recurrent-Selection (RCRS) Breeding Scheme

Previous studies have shown grain Zn content to be a quantitative trait controlled by many loci, like grain yield. Therefore, new breeding approaches are required to efficiently integrate grain Zn content into CIMMYT's breeding programs without massively increasing their scale. Selection theory indicates that simultaneous selection for several quantitative traits can be achieved in modestly scaled breeding programs with dramatically reduced cycle lengths, compared to the current norm of 5–6 years. Such rapid-cycle breeding plans—developed to complete a breeding cycle in 1–2 years—were developed and tested extensively and successfully in the 1980s and 1990s for some crops but have not been applied in wheat cultivar development. With the advent of genomic selection, rapid-cycle breeding has become even more attractive. Since RCRS is new to wheat breeding for cultivar development, we have advanced in the Toluca screenhouse and the F2-derived F3 individual spikes advanced as small plots in Toluca field season for disease phenotyping and also seed increase to have yield plots in Obregon.

A training population specifically for biofortification breeding have been generated during the next 2–3 years. Prediction models were developed, validated and used in the RCRS breeding pipeline to achieve simultaneous higher genetic gains for grain Zn and grain yield.

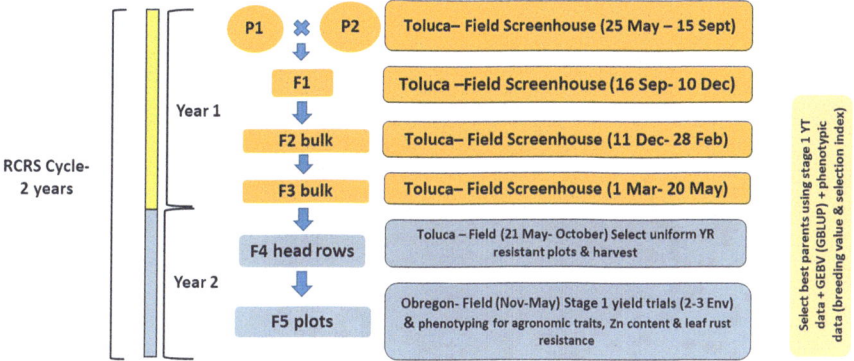

Fig. 4 Rapid cycle recurrent selection (RCRS) breeding scheme at CIMMYT initiated in 2021

Rapid-Cycling Recurrent Selection (RCRS) pipelines initiated with elite parents x high Zn parents and aiming to achieve 1 ppm annual Zn & 2% yield gain have been optimizied and running well. The RCRS scheme proposed in Fig. 4 is one is lesser than the RBGA scheme for generation advancement. Careful selection of parental lines from high Zn elite lines and highly elite germplasm from a non-biofortified pipeline, based on a quantitative estimate of breeding value using genotypic and limited phenotypic data along with selection indices calculated for each of the parental lines, is paramount to initiate 2nd breeding cycle in year 3.

Genomic Prediction Accuracies for Grain Yield in the Yield Testing Stage 1

The GBS data routinely used to predict the latest 1st year (stage 1) and Elite Yield Trials (EYTs, Stage 2). The stage 1 yield trials containing 5000 entries are phenotyped for grain yield in one environment, optimally irrigated to conduct selection for grain yield in diverse maturity groups. The prediction accuracies across YTs using environment, markers and pedigree model with all the available historic data in the training population has given varying results in different years and average around 0.3 (0.31 in Cycle 2013–2014, 0.20 in Cycle 2014–2015, 0.32 in Cycle 2015–2016, 0.18 in Cycle 2016–2017, 0.43 in Cycle 2017–2018 and 0.36 in Cycle 2018–2019). The GEBVs from this model have been used in making advancement decisions and results from over the years have shown that GEBVs can be best used in discarding lines vs selecting the best performers. For example in Fig. 5 below that shows the observed vs predicted grain yield for 8996 lines in Stage 1 yield trial 2018–2019 (grain yield predictions for the 8996 lines were done using a historic training set of ~50,000 lines evaluated during the past five cycles), we selected the top 10% of the lines using both phenotypic data and the genomic-estimated breeding values.

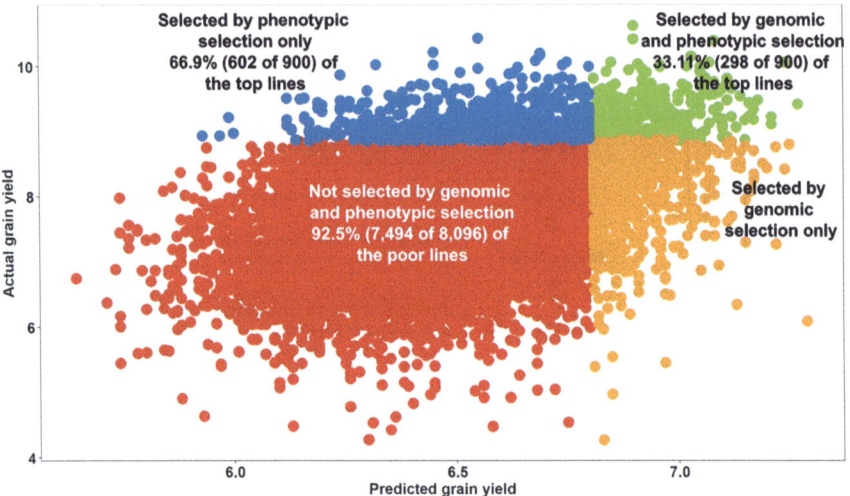

Fig. 5 Phenotypic and genomic selection for grain yield in Stage 1 Raised Bed 5IR environment, cycle 2018–2019

Among the 8096 lines with poor yield, GS was able to discard 92.5% of the lines (7494) that were also discarded by phenotypic selection. However, among the 900 top lines, GS was able to select only 33.11% of the top lines (298), leading to a risk of losing 66.9% of the top lines if we make selections based on the GEBVs alone.

Genomic Prediction Accuracies for Key Traits in the Yield Testing Stage 2

Genomic predictions have also been evaluated for EYTs (stage 2) with 1092 lines each and results for the 2014–2017 seasons provided here. In within-panel predictions, about 218 lines were predicted using 874 lines from the same panel/evaluation year and in across-panel predictions the 1092 lines were predicted using three other panels comprising 3276 lines. We used four panels of wheat breeding lines comprising CIMMYT's second-year EYTs evaluated during the crop seasons of 2014–2017 for genomic predictions. These lines were genotyped using GBS and from the set of 78,606 unfiltered markers, subsets of markers with less than 70%, 50% and 10% missing data were used for genomic predictions. The genomic coverage associated with the subsets clearly showed a decreasing trend towards the proximal centromeric regions with stringent filtering for missing data, and thereby served as ideal sets for evaluating the effect of genomic coverage on the predictabilities of traits (16,072 markers showed high coverage, 9285 markers had moderate coverage and 2253 markers showed low coverage). Genomic prediction accuracies were obtained with the genomic best linear unbiased prediction approach using fivefold cross-validations within each EYT panel, in which folds comprising 153–196 lines were predicted from four other folds comprising 613–784 lines, and predictions across EYT panels, for which 766–980 lines in a panel were predicted from three other panels comprising 2505–2719 lines.

The traits grain color, seedling and field resistance to stem rust, mixing time, alveograph W, flour sedimentation, loaf volume, flour protein content, grain protein content and thousand kernel weight had the highest genomic predictabilities (0.60–0.85), whereas all the other traits were moderately predictable in the cross-validations. In predictions across panels, the traits with high predictabilities were seedling resistance to several stem rust races, grain color, mixing time, alveograph W, field resistance to stem rust and flour sedimentation with an average decrease of only 0.07 ± 0.05 in accuracy from the corresponding cross-validation accuracies. Low across-panel predictabilities were observed for traits like grain yield and days to maturity in all environments, Septoria tritici blotch, Spot blotch, field resistance to yellow rust, etc. with an average decrease of 0.20 ± 0.06 from the corresponding cross-validation accuracies.

Zinc Wheat Variety Deployment

About 25 varieties of zinc wheat that have been officially released for production in eight countries, and which are adapted to numerous growing regions and climates. Over the past 5 years, the number of households growing zinc wheat in Pakistan has risen rapidly across the country from 218,000 in 2018 to over 4 million in 2024. HarvestPlus and its partners continue to drive demand to scale up zinc wheat, with the aim of reaching all corners of the country where people need it most. In India, more than ten biofortified varieties available for farmers cultivation and it is estimated that more than 9 million ha grown under biofortified wheat and supplying biofortified wheat for consumption of more than eight million people and it is expected to increase in the coming years. In Bangladesh, the blast resistant biofortified wheat BARI Gom 33 occupying more than 30% of wheat area in Bangladesh, this was due to the efforts made by the government policies and promotion activities. In Nepal, six biofortified wheat varieties grown by farmers and the expected acreage under Zn wheat expected to be around 20% of the total wheat area.

QTL Mapping Studies in Wheat Biofortification

One of the essential criteria for a successful breeding program is the availability of germplasm with large genetic diversity. Therefore, for wheat biofortification, several germplasm accessions have been examined and identified in primary, secondary and tertiary gene pools (for example in *T. aestivum*, *T. timopheevii*, *Secale cereale,* respectively) (Gupta et al., 2021). There is rich genetic variation for grain Fe and Zn concentrations from diverse wild tetraploid species such as *Triticum dicoccoides* (Zn conc. range: 20–159 ppm) and wild diploid species such as *Triticum boeoticum* (Fe conc. range: 41–92 ppm; Zn conc. range: 45–177 ppm) and *Triticum tauschii* (Fe conc. range: 59–99 ppm; Zn conc. range: 38–69 ppm) (Cakmak et al., 2000; Monasterio & Graham, 2000).

To date, several wheat genes that can contribute to Fe/Zn biofortification have been identified, and a major bottleneck for finding novel useful genes is due to their polyploidy, large size and high nucleotide similarity between different genome sets. Despite the rapid recent growth in the availability of information and tools for molecular genetic studies on wheat (Borrill et al., 2014), QTL studies remain important for the investigation of genes for wheat genetic biofortification. For quantitative traits such as grain zinc concentration (GZnC) and grain iron concentration (GFeC), QTL mapping studies are highly useful for discovering candidate genes and in the development of markers for marker-assisted breeding (Velu & Singh, 2019).

In a QTL analysis study by Tiwari et al. (2009), one QTL for GZnC on chromosome 7A explained a total phenotypic variance (PVE) of 18.8%, while two QTLs each for GFeC on chromosomes 2A and 7A explained 12.6% and 11.7% of the explained PVE respectively. Srinivasa et al. (2014) reliably identified in their study two QTLs for Zn and three QTLs for Fe across seven environments from a RIL population developed from a cross between *Triticum spelta* and *T. aestivum*. In another prominent study with three different sets of RILs by Crespo-Herrera et al. (2017), several QTLs were identified, of which two are of the highest interest, namely (i) *QGZn.cimmyt7B_1P2* on chromosome 7B that explained the major proportion of the total PVE (32.7%) for GZnC; and (ii) *QGFe. cimmyt-4A_P2* on chromosome 4A that explained a PVE of 21.14% for GFeC. In another study, where the RIL mapping population was derived from a cross between tetraploid and hexaploid wheat, Velu et al. (2017a) detected two major QTLs, namely 1B and 6B, for grain zinc concentration, and a third QTL on 2B that was also positively co-related to grain iron concentration, indicating that GZnC and GFeC can be selected together for biofortifying wheat. Since similar results of statistically significant positive correlation between GZnC and GFeC were observed in a study by Velu et al. (2011), it can be presumed that breeding simultaneously for Fe and Zn concentrations in wheat grain is feasible. A few other studies also suggest the simultaneous improvement of both mineral nutrient concentrations and seed size/TKW (thousand kernel weight).

In Silico Analysis

An *in silico* BLAST search of stable, high-PVE QTLs for GFeC and GZnC, and QTLs pleiotropic for GZnC and TKW, identified 37 candidate genes from the EnsemblPlants website, which are listed in Table 1. Some of the important putative candidate genes belonged to the cysteine synthase, beta glucanase, ion binding (zinc ion binding, iron ion binding, metal ion binding, magnesium ion binding), protein,

Table 1 In silico analysis of GFeC, GZnC QTLs, and QTLs pleiotropic for GZnC and TKW

QTL	Chr	Physical distance	Putative candidate gene	Description	Molecular function
GFeC					
QFeC.cimmyt-2B	2B	24,899,481–24,899,549	TraesCS2B02G050300 TraesCS2B02G050400	Stress responsive alpha-beta barrel, Dimeric alpha-beta barrel	–
			TraesCS2B02G050500	Leucine-rich repeat domain superfamily	–
			TraesCS2B02G050700	Peptidase S8/S53 domain superfamily, Subtilisin-like protease	Serine-type peptidase activity
			TraesCS2B02G050800	Cucumisin-like catalytic domain, Peptidase S8/S53 domain superfamily	Serine-type peptidase activity
			TraesCS2B02G050900	Non-haem dioxygenase N-terminal domain, Oxoglutarate/iron-dependent dioxygenase, Isopenicillin N synthase-like superfamily,	Metal ion binding, oxidoreductase activity,
			TraesCS2B02G051000	Cullin-associated NEDD8-dissociated protein 1/2, TATA-binding protein interacting (TIP20), Armadillo-type fold	SCF complex assembly
			TraesCS2B02G050200 TraesCS2B02G050100 TraesCS2B02G050000	Jacalin-like lectin domain superfamily, Allene oxide cyclase/Dirigent protein	Carbohydrate binding
			TraesCS2B02G049900	Cytochrome P450	Monooxygenase activity, iron ion binding, oxidoreductase activity, heme binding
QFeC.cimmyt-6A	6A	64,351,188–64,351,256	TraesCS6A02G097300	Domain of unknown function DUF1618	–
			TraesCS6A02G097100 TraesCS6A02G097200	CRAL-TRIO lipid binding domain	–
			TraesCS6A02G097400	Cytochrome P450	Monooxygenase activity, iron ion binding, oxidoreductase activity, heme binding

(continued)

Table 1 (continued)

QTL	Chr	Physical distance	Putative candidate gene	Description	Molecular function
QFeC.cimmyt-7A	7A	699,866,194–699,866,216	TraesCS7A02G512500	F-box-like domain superfamily	Protein binding
			TraesCS7A02G512700	Ribonuclease III, endonuclease domain superfamily, Double-stranded RNA-binding domain	Ribonuclease III activity, RNA binding
			TraesCS7A02G512300	Indole-3-glycerol phosphate synthase, Aldolase-type TIM barrel, Ribulose-phosphate binding barrel	Indole-3-glycerol-phosphate synthase activity, catalytic activity
			TraesCS7A02G512600	NADH-ubiquinone oxidoreductase chain 4L/K	Oxidoreductase activity, acting on NAD(P)H
GZnC					
QZnC.cimmyt-2B.3	2B	197,708,895–197,708,963	TraesCS2B02G213200	Formin, FH2 domain	Actin filament binding
			TraesCS2B02G213000	Fatty acid desaturase domain	Protein domain specific binding
			TraesCS2B02G213300	Pentatricopeptide repeat	Zinc ion binding
			TraesCS2B02G213400	Interferon-related developmental regulator	–
QZnC.cimmyt-7D	7D	58,828,355–58,828,382	TraesCS7D02G096800	Cysteine synthase	Cysteine synthase activity
			TraesCS7D02G096900 TraesCS7D02G097600	Histone H2A/H2B/H3	Structural constituent of chromatin
			TraesCS7D02G097000	Isoprenoid synthase domain superfamily	Magnesium ion binding
			TraesCS7D02G097100 TraesCS7D02G097200 TraesCS7D02G097500	Disease resistance protein (plants)	ADP binding
			TraesCS7D02G097300 TraesCS7D02G097400	Beta-glucanase	Hydrolase activity, hydrolyzing O-glycosyl compounds

GZnC & TKW					
QZnC.cimmyt-1D & QTKW.cimmyt-1D	1D	70,459,678–70,459,741	TraesCS1D02G085900	Phosphoglycerate/bisphosphoglycerate mutase, Histidine phosphatase superfamily, clade-1	Catalytic activity
			TraesCS1D02G086400	Zinc finger, CCHC-type, Nucleic acid-binding, OB-fold	Zinc ion binding, DNA binding
QZnC.cimmyt-2B.2 & QTKW.cimmyt-2B.1	2B	Un:238,403,796–238,403,828	TraesCSU02G196200	Sugar/inositol transporter, Major facilitator, sugar transporter-like, MFS transporter superfamily	Carbohydrate transport
			TraesCSU02G196100	CCT domain, Signal transduction response regulator, receiver domain, CheY-like superfamily	Protein binding
QZnC.cimmyt-3B.1 & QTKW.cimmyt-3B.5	3B	689,686,135–689,686,164	TraesCS3B02G449200	E3 ubiquitin-protein ligase RGLG, RING finger, plant	Ubiquitin protein ligase activity
			TraesCS3B02G449100	Amino acid transporter, transmembrane domain	Amino acid transport
			TraesCS3B02G448900	SANT/Myb domain, Homeobox-like domain superfamily	DNA binding

and ADP, RNA and DNA binding groups. Candidate genes with transmembrane transport, amino acid transport, molecular function, protein processing enzymes such as peptidases, and carbohydrate processing enzymes such as phosphoglycerate, sugar transporter genes and disease resistance genes were also identified.

Candidate Gene Analysis

The in silico BLAST search of the QTL-associated marker allele sequence identified several putative candidate genes for GZnC, GFeC, and pleiotropic for GZnC and TKW (Table 1). *QFeC.cimmyt-2B* region explaining 9.4% PVE appears to be linked to genes coding for Cytochrome P450, Cullin-associated NEDD8-dissociated protein, non-haem dioxygenase N-terminal domain, and Oxoglutarate/iron-dependent dioxygenase. Cytochrome P450 has a role in Fe and Zn regulation in Arabidopsis (van de Mortel et al., 2006) and iron ion binding. Cytochrome P450 is also presumed to be a putative candidate gene and is associated with QTLs governing micronutrient concentrations in earlier QTL mapping studies by Velu et al. (2017a, b), Liu et al. (2019), Cu et al. (2020), and Rathan et al. (2022). Non-heam dioxygenases could have a positive effect on improved yield and stress tolerance. Oxoglutarate/iron-dependent dioxygenase is involved in metal-ion bonding and oxidoreductase activity, which might play a role in iron homeostasis. Cullin-associated NEDD8-dissociated protein is a key assembly factor of SCF (SKP1-CUL1-F-box protein), E3 ubiquitin ligases, and acts as a F-box protein exchange factor. Also, Cytochrome P450 was found as a candidate gene near the *QFeC.cimmyt-6A* genomic region on chromosome 6A, which explained 8.3% PVE. Another QTL on chromosome 7A (*QFeC.cimmyt-7A*) that explained a PVE of 5.9% showed candidate gene coding for F-box-like domain superfamily and NADH-ubiquinone oxidoreductase chain 4L/K, where the former is associated with iron sensing and the latter is linked to oxidoreductase activity. F-box-like domain is also reported as a candidate gene in earlier studies (Krishnappa et al., 2022; Liu et al., 2019; Rathan et al., 2022). It is notable that FBXL5 (F-box and leucine-rich repeat protein 5) has an iron-responsive hemerythrin domain capable of binding iron, and an F-box domain that enables the ubiquitination of iron regulatory protein (IRP2) (Liang, 2022). FBXL5 protein increases in adequate iron conditions and vice-versa, thus acting as an iron sensor.

The QTL on chromosome 2B (*QZnC.cimmyt-2B.3*) genomic region with a PVE of 6.9% was found to be associated with putative genes for pentatricopeptide repeat. Interestingly, several pentatricopeptide repeats include C-terminal DYW deaminase domains which bind to zinc, as was evident in a study by Hayes et al. (2013). Remarkably, a pentatricopeptide repeat domain was found as a candidate gene for grain iron concentration in earlier studies by Rathan et al. (2021, 2022). Thus, it can be presumed that pentatricopeptide may have an increasing effect on both iron and zinc concentrations in wheat grain. *QZnC.cimmyt-7D* genomic region with a PVE of 17.6% was found to have important candidate genes coding for beta-glucanase and cysteine synthase. The β-Glucanase enzyme mediates the hydrolyzation of β-glucans, which are the sugars present in the aleurone cell walls of wheat bran and disintegrate the cell wall leading to the release of mineral nutrients (Yu et al., 2018). Cysteine formed as a product of cysteine synthase enzyme reaction is a sulphur-containing amino acid. It acts as a precursor of molecules such as nicotianamine which mediate Zn uptake, transport, and homeostasis (McDonald & Nik, 2009). Cysteines have a high affinity towards zinc ions (Zn^{2+}) and play a key role in protein structure, catalysis, and regulation (Pace & Weerapana, 2014). Zn-S bonds have a role in Zn release, Zn binding, control of Zn transfer reactions, cellular availability, and distribution of Zn, redox-active Zn proteins and Zn co-ordination dynamics (Maret, 2004).

The *QZnC.cimmyt-1D* genomic region is associated with a phosphatase enzyme that catalyses the phosphate ion and leads to the accumulation of Zn in grain. The role of phosphatase in increasing Zn concentration in grain has also been detected in previous studies (Rathan et al., 2022; Velu et al. 2017a, b, 2018). It is also noteworthy that phosphatase also increases the grain length, which indeed is positively correlated to TKW (Zhang et al. 2012; Velu et al. 2017a, b). Another candidate gene associated with this region was found to be zinc finger, CCHC type, which has a role in zinc ion binding and was also reported in a previous study (Rathan et al., 2022). Also, in a study by Velu et al. (2018), it is suggested that zinc finger motif and phosphatase enzyme play a crucial role in increasing Zn concentration in the wheat kernel. A candidate gene search in the *QZnC.cimmyt-2B.2* region detected a sugar/inositol transporter and CCT domain, wherein the latter is associated to protein binding and the former aids in carbohydrate transport, and hence is positively associated with TKW. A putative gene for *QZnC.cimmyt-3B.1* was identified to be E3 ubiquitin-protein ligase RGLG, RING finger. An E3 ubiquitin-protein ligase was also reported in a previous study by Cu et al. (2020) and is believed to be associated with increasing micronutrient concentrations in grain.

Conclusion

In summary, the wheat bioforification breeding program at CIMMYT have made great progress has been made in the past decade in transferring alleles for high-zinc (Zn) and iron (Fe) from diverse genetic resources into elite wheat breeding lines. However, the major challenge is to maintain simultaneous and high rates of genetic gains for grain yield and grain Zn to meet the food and nutritional security demands through the continuous delivery of biofortified varieties that are competitive to replace non-biofortified varieties successfully. Although a few intermediate effect QTL regions are identified for grain Zn, both yield and Zn content are quantitatively inherited. Increased breeding efforts and new approaches implemented to mainstream grain Zn and grain yield combine them in high frequency in CIMMYT's elite germplasm, ensuring that Zn levels are steadily increased to the required levels across the CIMMYT breeding pipelines. The addition of Zn as a core-trait achieved through significant acceleration in the breeding cycle, expanding population sizes, extensive Zn phenotyping, yield testing, phenotyping for biotic and abiotic stresses, molecular-assisted selection and genomic selection. While continuing to increase agronomic performance, high Zn alleles has been added as a core-trait. Eventually Zn content have been increased in the elite lines annually along with the frequency of elite lines with high yield and other agronomic traits that have potential to be released by partners. A genomics assisted "rapid cycle recurrent selection" scheme achieved through rapid generation advancement approaches enable CIMMYT wheat breeding program to mainstream grain Zn in the majority of elite lines.

References

2018 Global Nutrition Report. (2018). *Shining a light to spur action on nutrition*. Development Initiatives.

Balk, J., Connorton, J. M., Wan, Y., Lovegrove, A., Moore, K. L., Uauy, C., Sharp, P. A., & Shewry, P. R. (2019). Improving wheat as a source of iron and zinc for global nutrition. *Nutrition Bulletin, 44*(1), 53–59. https://doi.org/10.1111/nbu.12361

Bechtel, D. B., Abecassis, J., Shewry, P. R., & Evers, A. D. (2009). Development, structure, and mechanical properties of the wheat grain. In *Wheat: Chemistry and technology* (4th ed., pp. 51–95). AACC International, Inc.. https://doi.org/10.1016/B978-1-891127-55-7.50010-0

Borrill, P., Connorton, J., Balk, J., Miller, T., Sanders, D., & Uauy, C. (2014). Biofortification of wheat grain with iron and zinc: Integrating novel genomic resources and knowledge from model crops. *Frontiers in Plant Science, 5*. https://www.frontiersin.org/articles/10.3389/fpls.2014.00053

Bouis, H. E. (2003). Micronutrient fortification of plants through plant breeding: Can it improve nutrition in man at low cost? *The Proceedings of the Nutrition Society, 62*(2), 403–411. https://doi.org/10.1079/pns2003262

Bouis, H. E., & Welch, R. M. (2010). Biofortification—A sustainable agricultural strategy for reducing micronutrient malnutrition in the global south. *Crop Science, 50*(S1), S-20–S-32. https://doi.org/10.2135/cropsci2009.09.0531

Broadley, M. R., White, P. J., Hammond, J. P., Zelko, I., & Lux, A. (2007). Zinc in plants. *New Phytologist, 173*(4), 677–702. https://doi.org/10.1111/j.1469-8137.2007.01996.x

Cakmak, I., Ozkan, H., Braun, H. J., Welch, R. M., & Romheld, V. (2000). Zinc and iron concentrations in seeds of wild, primitive, and modern wheats. *Food and Nutrition Bulletin, 21*(4), 401–403. https://doi.org/10.1177/156482650002100411

Chasapis, C. T., Ntoupa, P.-S. A., Spiliopoulou, C. A., & Stefanidou, M. E. (2020). Recent aspects of the effects of zinc on human health. *Archives of Toxicology, 94*(5), 1443–1460. https://doi.org/10.1007/s00204-020-02702-9

Crespo-Herrera, L. A., Govindan, V., Stangoulis, J., Hao, Y., & Singh, R. P. (2017). QTL mapping of grain Zn and Fe concentrations in two hexaploid wheat RIL populations with ample transgressive segregation. *Frontiers in Plant Science, 8*. https://www.frontiersin.org/articles/10.3389/fpls.2017.01800

Cu, S. T., Guild, G., Nicolson, A., Velu, G., Singh, R., & Stangoulis, J. (2020). Genetic dissection of zinc, iron, copper, manganese and phosphorus in wheat (Triticum Aestivum L.) grain and rachis at two developmental stages. *Plant Science, 291*, 110338. https://doi.org/10.1016/j.plantsci.2019.110338

Erenstein, O., Jaleta, M., Mottaleb, K.A., Sonder, K., Donovan, J., Braun, HJ. (2022). Global Trends in Wheat Production, Consumption and Trade. In: Reynolds, M.P., Braun, HJ. (eds) Wheat Improvement. Springer, Cham. https://doi.org/10.1007/978-3-030-90673-3_4

FAO, IFAD, UNICEF, WFP, & WHO. (2021). Transforming food systems for food security, improved nutrition and affordable healthy diets for all. In *The state of food security and nutrition in the world 2021*. FAO. https://doi.org/10.4060/cb4474en

Fischer Walker, C. L., Ezzati, M., & Black, R. E. (2009). Global and regional child mortality and burden of disease attributable to zinc deficiency. *European Journal of Clinical Nutrition, 63*(5), 591–597. https://doi.org/10.1038/ejcn.2008.9

Gooding, M. J., & Shewry, P. R. (2022). The structure and composition of the wheat grain. In *Wheat* (pp. 263–300). Wiley. https://doi.org/10.1002/9781119652601.ch7

Gupta, S., Brazier, A. K. M., & Lowe, N. M. (2020). Zinc deficiency in low- and middle-income countries: Prevalence and approaches for mitigation. *Journal of Human Nutrition and Dietetics: The Official Journal of the British Dietetic Association, 33*(5), 624–643. https://doi.org/10.1111/jhn.12791

Gupta, P. K., Balyan, H. S., Sharma, S., & Kumar, R. (2021). Biofortification and bioavailability of Zn, Fe and Se in wheat: Present status and future prospects. *TAG. Theoretical and Applied Genetics. Theoretische Und Angewandte Genetik, 134*(1), 1–35. https://doi.org/10.1007/s00122-020-03709-7

Hayes, M. L., Giang, K., Berhane, B., & Michael Mulligan, R. (2013). Identification of two pentatricopeptide repeat genes required for RNA editing and zinc binding by C-terminal cytidine deaminase-like domains. *The Journal of Biological Chemistry, 288*(51), 36519–36529. https://doi.org/10.1074/jbc.M113.485755

Hazard, B., Trafford, K., Lovegrove, A., Griffiths, S., Uauy, C., & Shewry, P. (2020). Strategies to improve wheat for human health. *Nature Food, 1*(8), 475–480. https://doi.org/10.1038/s43016-020-0134-6

Hussain, S., Maqsood, M., Rengel, Z., Aziz, T., & Abid, M. (2013). Estimated zinc bioavailability in milling fractions of biofortified wheat grains and in flours of different extraction rates. *International Journal of Agriculture and Biology, 15*, 383–388.

Institute of Medicine. (2001). *Dietary reference intakes for vitamin A, vitamin K, arsenic, boron, chromium, copper, iodine, iron, manganese, molybdenum, nickel, silicon, vanadium, and zinc*. National Academies Press. https://doi.org/10.17226/10026

International Wheat Genome Sequencing Consortium (IWGSC). (2014). A chromosome-based draft sequence of the hexaploid bread wheat (Triticum aestivum) genome. *Science (New York, N.Y.), 345*(6194), 1251788. https://doi.org/10.1126/science.1251788

International Zinc Nutrition Consultative Group (IZiNCG), Brown, K. H., Rivera, J. A., Bhutta, Z., Gibson, R. S., King, J. C., Lönnerdal, B., et al. (2004). International Zinc Nutrition Consultative Group (IZiNCG) technical document #1. Assessment of the risk of zinc deficiency in populations and options for its control. *Food and Nutrition Bulletin, 25*(1 Suppl 2), S99–S203.

Krishnappa, G., Khan, H., Krishna, H., Kumar, S., Mishra, C. N., Parkash, O., Devate, N. B., et al. (2022). Genetic dissection of grain iron and zinc, and thousand kernel weight in wheat (Triticum Aestivum L.) using genome-wide association study. *Scientific Reports, 12*(1), 12444. https://doi.org/10.1038/s41598-022-15992-z

Liang, G. (2022). Iron uptake, signaling, and sensing in plants. *Plant Communications*, 100349. https://doi.org/10.1016/j.xplc.2022.100349

Liu, J., Bihua, W., Singh, R. P., & Velu, G. (2019). QTL mapping for micronutrients concentration and yield component traits in a hexaploid wheat mapping population. *Journal of Cereal Science, 88*, 57–64. https://doi.org/10.1016/j.jcs.2019.05.008

Lockyer, S., White, A., & Buttriss, J. L. (2018). Biofortified crops for tackling micronutrient deficiencies – What impact are these having in developing countries and could they be of relevance within Europe? *Nutrition Bulletin, 43*(4), 319–357. https://doi.org/10.1111/nbu.12347

Maret, W. (2004). Zinc and sulfur: A critical biological partnership. *Biochemistry, 43*(12), 3301–3309. https://doi.org/10.1021/bi036340p

Maxfield, L., Shukla, S., & Crane, J. S. (2022). Zinc deficiency. In *StatPearls [Internet]*. StatPearls Publishing. https://www.ncbi.nlm.nih.gov/books/NBK493231/

McDonald, G. K., & Nik, M. M. (2009, April). *Increasing the supply of sulphur increases the grain zinc concentration in bread and durum wheat.* https://escholarship.org/uc/item/43k2r1h8

Monasterio, I., & Graham, R. D. (2000). Breeding for trace minerals in wheat. *Food and Nutrition Bulletin, 21*(4), 392–396. https://doi.org/10.1177/156482650002100409

Nestel, P., Bouis, H. E., Meenakshi, J. V., & Pfeiffer, W. (2006). Biofortification of staple food crops. *The Journal of Nutrition, 136*(4), 1064–1067. https://doi.org/10.1093/jn/136.4.1064

OECD. (2021). *OECD-FAO agricultural outlook (edition 2021).* Organisation for Economic Co-operation and Development. https://www.oecd-ilibrary.org/agriculture-and-food/data/oecd-agriculture-statistics/oecd-fao-agricultural-outlook-edition-2021_4bde2d83-en?parentId =http%3A%2F%2Finstance.metastore.ingenta.com%2Fcontent%2Fcollection%2Fagr-data-en

Pace, N. J., & Weerapana, E. (2014). Zinc-binding cysteines: Diverse functions and structural motifs. *Biomolecules, 4*(2), 419–434. https://doi.org/10.3390/biom4020419

Pfeiffer, W. H., & McClafferty, B. (2007). HarvestPlus: Breeding crops for better nutrition. *Crop Science, 47*(S3), S-88–S-105. https://doi.org/10.2135/cropsci2007.09.0020IPBS

Rathan, N. D., Sehgal, D., Thiyagarajan, K., Singh, R., Singh, A.-M., & Govindan, V. (2021). Identification of genetic loci and candidate genes related to grain zinc and iron concentration using a zinc-enriched wheat 'zinc-shakti'. *Frontiers in Genetics, 12*, 652653. https://doi.org/10.3389/fgene.2021.652653

Rathan, N. D., Krishna, H., Ellur, R. K., Sehgal, D., Govindan, V., Ahlawat, A. K., Krishnappa, G., et al. (2022). Genome-wide association study identifies loci and candidate genes for grain micronutrients and quality traits in wheat (Triticum Aestivum L.). *Scientific Reports, 12*(1), 7037. https://doi.org/10.1038/s41598-022-10618-w

Roohani, N., Hurrell, R., Kelishadi, R., & Schulin, R. (2013). Zinc and its importance for human health: An integrative review. *Journal of Research in Medical Sciences: The Official Journal of Isfahan University of Medical Sciences, 18*(2), 144–157.

Rosado, J. L., Michael Hambidge, K., Miller, L. V., Garcia, O. P., Westcott, J., Gonzalez, K., Conde, J., et al. (2009). The quantity of zinc absorbed from wheat in adult women is enhanced by biofortification. *The Journal of Nutrition, 139*(10), 1920–1925. https://doi.org/10.3945/jn.109.107755

Singh, R., Velu, G., Andersson, M., Bouis, H., & Jamora, N. (2017). *Zinc-biofortified wheat: Harnessing genetic diversity for improved nutritional quality.* CIMMYT, HarvestPlus, and the Global Crop Diversity Trust.

Srinivasa, J., Arun, B., Mishra, V. K., Singh, G. P., Velu, G., Babu, R., Vasistha, N. K., & Joshi, A. K. (2014). Zinc and iron concentration QTL mapped in a Triticum spelta × T. aestivum cross. *TAG. Theoretical and Applied Genetics. Theoretische Und Angewandte Genetik, 127*(7), 1643–1651. https://doi.org/10.1007/s00122-014-2327-6

Tiwari, V. K., Rawat, N., Chhuneja, P., Neelam, K., Aggarwal, R., Randhawa, G. S., Dhaliwal, H. S., Keller, B., & Singh, K. (2009). Mapping of quantitative trait loci for grain iron and zinc concentration in diploid a genome wheat. *The Journal of Heredity, 100*(6), 771–776. https://doi.org/10.1093/jhered/esp030

van de Mortel, J. E., Villanueva, L. A., Schat, H., Kwekkeboom, J., Coughlan, S., Moerland, P. D., van Themaat, E. V. L., Koornneef, M., & Aarts, M. G. M. (2006). Large expression differences in genes for iron and zinc homeostasis, stress response, and lignin biosynthesis distinguish roots of *Arabidopsis thaliana* and the related metal hyperaccumulator *Thlaspi caerulescens*. *Plant Physiology, 142*(3), 1127–1147. https://doi.org/10.1104/pp.106.082073

Velu, G., & Singh, R. P. (2019). Genomic approaches for biofortification of grain zinc and iron in wheat. In A. M. Iqbal Qureshi, Z. A. Dar, & S. H. Wani (Eds.), *Quality breeding in field crops* (pp. 193–198). Springer International Publishing. https://doi.org/10.1007/978-3-030-04609-5_9

Velu, G., Singh, R., Huerta-Espino, J., Peña, J., & Ortiz-Monasterio, I. (2011). Breeding for enhanced zinc and iron concentration in CIMMYT spring wheat germplasm. *Czech Journal of Genetics and Plant Breeding, 47*(Special Issue), S174–S177. https://doi.org/10.17221/3275-CJGPB

Velu, G., Ortiz-Monasterio, I., Cakmak, I., Hao, Y., & Singh, R. P. (2014). Biofortification strategies to increase grain zinc and iron concentrations in wheat. *Journal of Cereal Science,*

Cereal Science for Food Security, Nutrition and Sustainability, 59(3), 365–372. https://doi.org/10.1016/j.jcs.2013.09.001

Velu, G., Singh, R. P., Huerta, J., & Guzmán, C. (2017a). Genetic impact of Rht dwarfing genes on grain micronutrients concentration in wheat. *Field Crops Research, 214*, 373–377. https://doi.org/10.1016/j.fcr.2017.09.030

Velu, G., Tutus, Y., Gomez-Becerra, H. F., Hao, Y., Demir, L., Kara, R., Crespo-Herrera, L. A., et al. (2017b). QTL mapping for grain zinc and iron concentrations and zinc efficiency in a tetraploid and hexaploid wheat mapping populations. *Plant and Soil, 411*(1), 81–99. https://doi.org/10.1007/s11104-016-3025-8

Velu, G., Singh, R. P., Crespo-Herrera, L., Juliana, P., Dreisigacker, S., Valluru, R., Stangoulis, J., et al. (2018). Genetic dissection of grain zinc concentration in spring wheat for mainstreaming biofortification in CIMMYT wheat breeding. *Scientific Reports, 8*, 13526. https://doi.org/10.1038/s41598-018-31951-z

Velu, G., Singh, R. P., & Joshi, A. K. (2020). 5 – A decade of progress on genetic enhancement of grain zinc and iron in CIMMYT wheat germplasm. In O. P. Gupta, V. Pandey, S. Narwal, P. Sharma, S. Ram, & G. P. Singh (Eds.), *Wheat and barley grain biofortification* (pp. 129–138). Woodhead Publishing Series in Food Science, Technology and Nutrition. Woodhead Publishing. https://doi.org/10.1016/B978-0-12-818444-8.00005-5

Velu, G., Michaux, K. D., & Pfeiffer, W. H. (2022). Nutritionally enhanced wheat for food and nutrition security. In M. P. Reynolds & H.-J. Braun (Eds.), *Wheat improvement: Food security in a changing climate* (pp. 195–214). Springer International Publishing. https://doi.org/10.1007/978-3-030-90673-3_12

Virk, P. S., Andersson, M. S., Arcos, J., Govindaraj, M., & Pfeiffer, W. H. (2021). Transition from targeted breeding to mainstreaming of biofortification traits in crop improvement programs. *Frontiers in Plant Science, 12*. https://www.frontiersin.org/articles/10.3389/fpls.2021.703990

Wan, Y., Stewart, T., Amrahli, M., Evans, J., Sharp, P., Govindan, V., Hawkesford, M. J., & Shewry, P. R. (2022). Localisation of iron and zinc in grain of biofortified wheat. *Journal of Cereal Science, 105*, 103470. https://doi.org/10.1016/j.jcs.2022.103470

Wessells, K. R., & Brown, K. H. (2012). Estimating the global prevalence of zinc deficiency: Results based on zinc availability in national food supplies and the prevalence of stunting. *PLoS One, 7*(11), e50568. https://doi.org/10.1371/journal.pone.0050568

WHO. (2021). *Fact sheets – Malnutrition*. https://www.who.int/news-room/fact-sheets/detail/malnutrition

Yu, X., Han, J., Li, H., Zhang, Y., & Feng, J. (2018). The effect of enzymes on release of trace elements in feedstuffs based on in vitro digestion model for monogastric livestock. *Journal of Animal Science and Biotechnology, 9*(1), 73. https://doi.org/10.1186/s40104-018-0289-2

Zhang, Y.-Q., Sun, Y.-X., You-Liang Ye, M., Karim, R., Xue, Y.-F., Yan, P., Meng, Q.-F., et al. (2012). Zinc biofortification of wheat through fertilizer applications in different locations of China. *Field Crops Research, 125*, 1–7. https://doi.org/10.1016/j.fcr.2011.08.003

Open Access This chapter is licensed under the terms of the Creative Commons Attribution 4.0 International License (http://creativecommons.org/licenses/by/4.0/), which permits use, sharing, adaptation, distribution and reproduction in any medium or format, as long as you give appropriate credit to the original author(s) and the source, provide a link to the Creative Commons license and indicate if changes were made.

The images or other third party material in this chapter are included in the chapter's Creative Commons license, unless indicated otherwise in a credit line to the material. If material is not included in the chapter's Creative Commons license and your intended use is not permitted by statutory regulation or exceeds the permitted use, you will need to obtain permission directly from the copyright holder.

Breeding and Deploying High-Zinc Maize in the Tropics

Prasanna Boddupalli, Natalia Palacios-Rojas, Felix San Vicente, Thanda Dhliwayo, Abebe Menkir, Thokozile Ndhlela, Sudha K. Nair, and Xuecai Zhang

Introduction

Maize (*Zea mays* L,) is a major staple food in Sub-Saharan Africa (SSA) and Latin America, and in some countries in Asia. It is the most important energy source in SSA, where consumption ranges from 50 to >330 g/person/day, while in Latin America, the same ranges from 50 to >300 g/person/day (Prasanna et al., 2020). Besides, maize is a source of diverse micronutrients and phytochemicals, such as phenolics, carotenoids (yellow and orange maize), anthocyanins (blue, purple, and black maize), phlobaphenes (red maize), insoluble and soluble dietary fiber, and polar and non-polar lipids, hence all other compounds providing health benefits to humans. The concentrations of various nutrients in maize kernels depend on the genetic background, agronomic management, interaction between the genotype and the environment, and post-harvest handling (Ekpa et al., 2019).

Elevated carbon dioxide (eCO_2) levels, resulting from environmental factors, have led to accelerated plant growth rates. However, this phenomenon has a detrimental effect on plant protein content and essential micronutrients, such as calcium,

---

P. Boddupalli (✉)
International Maize and Wheat Improvement Center (CIMMYT), Nairobi, Kenya
e-mail: b.m.prasanna@cgiar.org

N. Palacios-Rojas · F. S. Vicente · T. Dhliwayo · X. Zhang
International Maize and Wheat Improvement Center (CIMMYT), Texcoco, Estado de México, Mexico

A. Menkir
International Institute on Tropical Agriculture (IITA), Ibadan, Nigeria

T. Ndhlela
CIMMYT, Harare, Zimbabwe

S. K. Nair
CIMMYT, ICRISAT Campus, Patancheru, Greater Hyderabad, Telangana, India

© The Author(s) 2025
M. Govindaraj et al. (eds.), *Breeding Zinc Crops for Better Human Health*, https://doi.org/10.1007/978-3-031-84342-6_3

iron, and zinc. Many studies have shown that the grains and tubers of C3 plants, including staples like wheat, rice, and barley, can experience reductions of up to 15% in protein content and up to 11% in these vital minerals (Myers et al., 2014; Smith & Myers, 2018; Marcos-Barbero et al., 2021). These crops, along with maize, constitute the dietary foundation for billions of people worldwide, underscoring the critical role of their mineral content in human health. While maize, a C4 plant, exhibits better adaptability to the decreasing zinc levels caused by eCO_2, the zinc levels in maize kernels are not sufficiently high for substantial nutritional impact.

Intensive agricultural practices have introduced risks to soil health, including erosion, nutrient imbalances, and the depletion of organic matter. Moreover, climate change presents additional challenges related to soil fertility, such as altered precipitation patterns, increased temperatures, and shifting weather conditions, all of which contribute to soil degradation and reduced nutrient availability. These factors exacerbate the nutrient content decline in staple cereals and legumes, as these crops heavily depend on healthy soils for optimal growth (Hummel et al., 2018). According to Owino et al. (2022), there will be a 13.6% decrease in soil iron and a 14.6% decrease in soil zinc by 2050, factoring in the combined effects of projected atmospheric CO_2 increases, including the carbon nutrient penalty, CO_2 fertilisation, and climate-related productivity changes.

Zinc (Zn), a micronutrient essential for human health and well-being, plays a very important role in numerous enzymatic processes, immune system function, growth, and development. Zinc plays a critical role in immune function and may have a role in preventing viral infections, including COVID-19. Although not a standalone solution, Zn helps modulate the immune response, has direct antiviral properties that can limit viral replication, strengthen mucosal barriers, support immune cell function, and help regulate inflammation (Mayor-Ibarguren & Robles-Marhuenda, 2020). Apart from its role in immune system function, Zn is one of the basic nutrients required during pregnancy for the normal development and growth of the fetus. However, Zn deficiency during pregnancy causes birth outcomes such as growth impairment, spontaneous abortion, congenital malformations, stillbirths, and preterm births, among others (Agedew et al., 2022). Zn is mostly stored in the skeletal muscles, and the total amount of zinc in an adult human body is approximately 2–3 g, with a daily requirement of 12–16 mg. Decreased intake, malabsorption, or high losses of micronutrients from the gut may lead to Zn deficiency, resulting in impaired growth, neuronal abnormalities, iron deficiency anemia due to decreased iron absorption, and even cardiovascular diseases (Ahsan et al., 2021).

Despite its critical significance, Zn deficiency remains a pressing global health concern, particularly in regions of the Global South where maize is not just a staple but an integral part of food culture, as seen in Southern Africa, Mexico, Mesoamerica, and some countries in South America (Brown et al., 2004; Caulfield & Black, 2004; Lopez-Ridaura et al., 2021; Birol & Bouis, 2023). In young children, Zn deficiency increases the risk of diarrhea, pneumonia, malaria, and mortality from those diseases. Zn deficiency was reported to result in more than 0.5 million deaths per year in infants and children below 5 years of age (Krebs et al., 2014). With roughly 17% of the global population consuming a zinc-deficient diet, around a quarter of the pediatric population below the age of 5 suffers from stunted growth (de Benoist et al., 2007). Several measures have been suggested to treat symptoms and prevent zinc deficiency, including recommendations by the WHO and UNICEF to administer 20 mg oral zinc supplements for 10–14 days in children suffering from diarrhea to help them recover micronutrient losses, as zinc supplements reduce deaths from diarrhea and pneumonia in children by 13% and 15%, respectively (Ahsan et al., 2021).

Biofortification is another potential strategy to alleviate micronutrient malnutrition with the intervention of genetic basis for improvement of major staple food crops like maize (Prasanna et al., 2020). By breeding improved maize germplasm with higher kernel Zn content, biofortification ensures efficient delivery of this essential nutrient at the required levels to those who need it the most. Furthermore, maize's deep-rooted cultural and culinary significance in many Global South communities makes biofortified maize varieties that retain traditional taste, texture, and appearance more likely to be embraced and consumed (CAST, 2020; Palacios-Rojas et al., 2020).

Crop and dietary diversification are also promising agronomic approaches for alleviating micronutrient deficiency and complementing biofortification. However, in the context of smallholder farmers, once biofortified maize varieties are integrated into the farming systems, they can provide increased zinc content for generations without much need for additional interventions. Knowing the most important factors vis-à-vis weather, soil, and crop management (e.g., fertilizer applied, residue management, etc.) will contribute to making the best use of biofortified crops and could help provide a more realistic evaluation of their possible effect on the nutrition and health of consumers (de Valença et al., 2017). Recent analyses suggested that biofortification could be a cost-effective approach—sometimes more cost-effective than alternative or complementary micronutrient interventions, including mineral fertilization and kitchen gardens, and even more so than many other public health or nutrition interventions like supplementation and industrial fortification (CAST, 2020).

Breeding High-Zinc Maize Germplasm

Genetic Variability for Kernel Zn Concentration in Maize Germplasm

Genetic variability for a particular trait is a prerequisite for making genetic improvements through breeding. Breeding for enhanced kernel Zn content in maize is challenging due to the heavy reliance of plants on soil mineral concentrations. Nutritional breeding targets for kernel Zn were established based on the amount of maize consumed (g day^{-1}), bioavailability (% Zn absorbed), retention after processing (e.g., milling, storage, and cooking), and the percentage of the daily requirement of Zn from maize (Akhtar et al., 2018; Bouis et al., 2011). In the case of maize, the minimum target level of kernel Zn (33 µg/g) is intended to provide at least 50% of the daily physiological requirement, assuming absorption and retention rates of 25% and 80%, respectively, upon consuming ~300 g of uncooked maize per day (Bouis et al., 2011). The average Zn concentration in maize ranges between 20 and 25 µg/g (Bouis et al., 2011; Chomba et al., 2015). According to recent data from Evangelina Villegas (Maize Quality Laboratory at CIMMYT, Mexico), further insights into the Zn content of grain samples from maize hybrids and landraces sourced from diverse regions. In Zimbabwe, analysis of 344 samples showed an average kernel Zn content of 18.6 µg/g (ranging from 11.1 to 30.7 µg/g). Hybrid maize grain samples from different regions in Mexico recorded an average kernel Zn concentration of 18.9 µg/g, with a range of 14.1–26.4 µg/g. In contrast, analysis of grain samples from 225 maize landraces from different regions in Mexico showed an average Zn content of 22.7 µg/g, with a broader range of 15.8–41.3 µg/g. These variations in Zn content can be attributed not only to the genetic diversity of the maize varieties but also to the specific locations and the agricultural practices employed by farmers. Overall, the data suggests that 8–13 µg/g improvements in kernel Zn content can be achieved in maize through conventional breeding.

Evaluating diverse germplasm for the presence of genetic variability is essentially the first step while breeding maize for improved kernel Zn concentration (Cakmak, 2008; Menkir, 2008). The objectives for evaluating the available germplasm are to identify: (i) parental genotypes that can be used in crosses, genetic studies, and molecular marker development; and (ii) existing varieties that combine high-Zn with desired agronomic traits for commercialization (Bouis & Saltzman, 2017). Several studies have reported the existence of wide genetic variation for kernel Zn concentration in different genetic backgrounds (Table 1), suggesting the potential of genetically increasing Zn content in maize. Screening 1400 maize genotypes and 400 landraces maintained in the gene bank of the International Maize and Wheat Improvement Center (CIMMYT, Mexico) showed significant genetic variation for kernel Zn content (Bänziger & Long, 2000). Menkir (2008) reported that among tropically adapted maize inbred lines of the International Institute of Tropical Agriculture (IITA, Nigeria), the best inbred lines possessed 14–180% higher grain Zn over the trial mean values. It is also reported in various studies that Zn contents in the grains of modern maize genotypes are lower than landraces due to yield dilution effects associated with higher grain yield in modern cultivars, larger germ size in landraces, or the quality of soil on which landraces are produced (McDonald et al., 2008; Pfeiffer & McClafferty, 2007a, b). Therefore, grain yield level should also be a key consideration while breeding for improved Zn contents in cereals to ensure acceptability by farmers (Garcia-Oliveira et al., 2018).

Besides understanding the extent of genetic variability for kernel Zn, several studies have explored the genetic and molecular bases of high Zn concentrations in relation to other components that could potentially affect or contribute to kernel Zn concentration. This knowledge is helpful for breeders, especially while developing selection criteria for improving kernel Zn (Chakraborti et al., 2009a, b, 2011a, b).

Table 1 Genetic variability for kernel Zn concentrations in maize reported in various studies since 2000

S.No.	Range Zn (µg/g)	Type of germplasm	No of entries	Country of evaluation	References
1	12.9–57.6	Landraces and improved genotypes	1814	Zimbabwe & Mexico	Bänziger and Long (2000)
2	11.6–95.6	Inbred lines	109	Nigeria	Maziya-Dixon et al. (2000)
3	16.0–23.6	Hybrids	28	Croatia	Brkic et al. (2004)
4	16.5–20.5	Varieties	20	Nigeria	Oikeh et al. (2003a)
5	16.5–24.6	Varieties	49	Nigeria	Oikeh et al. (2003b)
6	18.1–29.8	Inbreds	14	Zimbabwe	Long et al. (2004)
7	15.0–47.0	Landraces	400	Mexico	Ortiz-Monasterio et al. (2007)
8	14.0–45.0	Inbreds	310	Nigeria	Menkir (2008)
9	13.4–46.4	Inbreds	25	India	Chakraborti et al. (2009b)
10	16.4–28.6	F4 families	294	Croatia	Šimić et al. (2009)
11	17.6–49.1	Hybrids	49	India	Chakraborti et al. (2011a)
12	21.9–40.9	Inbreds	31	India	Chakraborti et al. (2011b)
13	19.3–30.9	Hybrids	42	Mexico & Ethiopia	Pixley et al. (2011)
14	15.1–53.0	Inbreds and landraces	30	India	Prasanna et al. (2011)
15	17.5–42.0	Inbreds	22	Brazil	Queiroz et al. (2011)
16	7.0–29.9	Inbreds and landraces	67	India	Agrawal et al. (2012)
17	3.8–35.8	Inbreds and landraces	81	India	Guleria et al. (2013)
18	12.6–39.4	QPM inbreds	46	India	Pandey et al. (2015)
19	16.4–53.2	Inbreds	188	India	Mallikarjuna et al. (2014)
20	19.4–32.6	Improved genotypes	48	India	Thakur et al. (2015)
21	20.0–53.0	Inbreds	24	Nigeria	Akinwale and Adewopo (2016)
22	17.1–43.8	Inbreds	923	Mexico	Hindu et al. (2018)
23	14.1–26.4	Hybrids	93	Mexico	CIMMYT (unpublished)
24	15.8–41.3	Landraces	225	Mexico	CIMMYT (unpublished)
25	11.1–30.7	Hybrids	344	Zimbabwe	CIMMYT (unpublished)
26	10.7–57.8	Hybrids	77	Zimbabwe	Goredema-Matongera et al. (2023)

Phenotyping for Kernel Zn Trait

Various factors can introduce variability in the results of studies investigating kernel Zn concentration in maize, including germplasm differences and environmental influences. The choice of quantification method, postharvest sample handling, and microenvironmental variations can all play a role (Prasanna et al., 2020). To ensure robust and accurate results, spectroscopic methods like Inductively Coupled Plasma Optical Emission Spectroscopy (ICP-OES) and Atomic Absorption Spectroscopy (AAS) are commonly employed. These methods, either through chemical or microwave-assisted sample digestion, offer high-throughput analysis. For large-scale screening, especially at the early stages of high-Zn maize breeding or molecular studies involving diverse samples, locations, and replicates, X-ray fluorescence (XRF) is widely used due to its various benefits, especially minimal sample preparation and no use of hazardous chemicals (Mageto et al., 2020a, b; Guo et al., 2020; Hindu et al., 2018).

Relationship between QPM and Kernel Zn

Early efforts to assess the potential for Zn biofortification in maize, including micronutrient surveys, found higher levels of Zn in Quality Protein Maize (QPM) genotypes than in the normal maize germplasm (Chakraborti et al., 2009a, b, 2011a, b; Hindu et al., 2018). These findings were consistent with the results of an earlier study showing increased levels of Zn, Ca, Mg, and Cu in inbred genotypes possessing the *opaque2* (*o2*) and the multiple aleurone layer (Mal) genes compared to the wild-type versions of the inbreds (Welch et al., 1993). The dominant, wild-type *O2* allele codes for a transcription factor that regulates the synthesis of zein proteins (Schmidt et al., 1990). The *o2* recessive allele results in a decrease in α-zein, with a proportional increase in non-zein proteins, including albumins, glutelins, and globulins (Habben et al., 1993). Non-zein proteins were shown to bind Zn in the endosperm of maize kernels (Diez-Altares & Bornemisza, 1967). Consequently, the elevated levels of Zn in QPM could be attributed to reduced levels of zeins and relatively higher levels of other Zn-binding proteins.

Significant genetic variability for kernel Zn has been reported not only in QPM backgrounds but also in normal maize germplasm (Agrawal et al., 2012; Long et al., 2004; Maziya-Dixon et al., 2000; Menkir, 2008). The CIMMYT maize breeding program has developed high-Zn inbreds and hybrids from both QPM and non-QPM germplasm sources (Mageto et al., 2020a, b; Gallego-Castillo et al., 2021). In a study to evaluate the combining ability for Zn content in maize kernels, Mageto et al. (2020a, b) reported the highest kernel Zn content in hybrids between high-Zn QPM and high-Zn normal inbred lines and the lowest between low-Zn QPM and low-Zn normal inbred lines, suggesting that for hybrid development, having high-Zn in the parents is more important than having the o2 recessive allele. The presence of genetic variation for kernel Zn in both QPM and non-QPM maize germplasm results in a broad germplasm pool that breeders can exploit to develop high-Zn and high-yielding varieties that meet consumer preferences in the target markets.

Breeding Methodology for Enriching Kernel Zn

Enriching maize with high Zn involves systematic evaluation of the extent of genetic diversity available for breeding, generating source populations from parents with high Zn content to extract Zn-enriched inbreds and select for other desirable agronomic traits, evaluating new inbred lines for Zn concentration at S4 or subsequent inbreeding stages, crossing the selected high-Zn lines with suitable testers, and evaluating the testcrosses for their agronomic performance. In addition, implementing strategies such as genomic selection and prediction can significantly lower costs, enhance efficiency, and accelerate the breeding process (Bouis & Saltzman, 2017).

Investigations on the genetics of the kernel Zn content of maize were first reported in the 1960s and 1970s (Gorsline et al., 1964; Arnold & Bauman, 1976). Additive gene action was found to be more important than non-additive gene action in the inheritance of kernel Zn concentration in maize (Gorsline et al., 1964; Arnold et al., 1977; Brkic et al., 2004; Long et al., 2004), indicating the potential that exists for realization of genetic gains from selection. Subsequent studies conducted in the 2000s focused on the assessment of the genetic potential of maize germplasm for increasing the concentration of Zn in maize grains (Bänziger & Long, 2000; Long et al., 2004; Oikeh et al., 2003a, b; Šimić et al., 2009). Genetic studies also revealed that kernel Zn concentration in maize is a quantitative trait with complex inheritance (Long et al., 2004; Baxter et al., 2013; Šimić et al., 2012). The complexity of the trait is accentuated by low heritability and higher environmental and genotype × environment interaction (GEI) effects. Despite these challenges, there is still huge potential to develop maize varieties with enhanced kernel Zn concentration.

Zinc biofortification of maize has been undertaken at CIMMYT and IITA, in partnership with public sector institutions. Substantial genetic variation for kernel-Zn (4–96 ppm) was found in tropical maize germplasm (Table 1), including landraces, inbred lines, hybrids, and open-pollinated varieties (OPVs). Unlike crops like wheat and rice (Guzmán et al., 2014), no significant correlation was observed between kernel Zn and Fe contents in maize. In fact, high Zn maize normally contains between 18 and 20 ppm Fe, which is the average content in maize kernel.

Breeding efforts on high-Zn maize were initially focused on Latin American countries, including Guatemala, Nicaragua, Honduras, and Colombia, and Western African countries, including Ghana, Benin, and Nigeria. In recent years, additional efforts on high-Zn maize breeding, leveraging on the ones being made in Latin America, have been initiated in southern African countries, particularly Zimbabwe, and South Asian countries, including India, Pakistan, and Nepal. High-Zn maize breeding at CIMMYT was mostly targeted at enriching white maize germplasm in both QPM and non-QPM genetic backgrounds. Three QPM CIMMYT Maize Lines (CMLs)—CML176, CML491, and CML492—were found to be particularly important for improving kernel Zn in tropical maize and have been used extensively as founder lines in pedigree-based selection. These elite QPM lines were derived from CIMMYT white flint QPM Population 62 and white dent QPM Population 63, which were subjected to several cycles of intra-population recurrent selection in the 1980s (CIMMYT, 1998). Population 62 traces back to the ETO composite, whose main components were the tropical Colombian landraces Comun and Chococeño and the Venezuelan landraces Puya and Cubano Amarillo. Population 63 traces back to the Tuxpeño-1 composite, whose main component was the Mexican landrace Tuxpeño (CIMMYT, 1998).

Adapted yellow and white endosperm maize inbred lines derived from broad-based populations, biparental crosses, and backcrosses with high Zn content have been used for generating pedigree populations at CIMMYT and IITA to develop new high-Zn inbred lines. Interestingly, an above-average concentration of kernel Zn was found in the QPM germplasm as compared to non-QPM/normal maize germplasm (Chakraborti et al., 2009a, b, 2011a, b). However, not all QPM germplasm is high in kernel Zn, and it is also possible to have non-QPM germplasm with high kernel Zn. Using 923 lines to conduct genome-wide association studies (GWAS) for kernel Zn, Hindu et al. (2018) reported that only 31 were QPM or had QPM background and 33.3% had Zn values higher than 30 μg g^{-1} on dry weight (DW) basis. In contrast, out of the 892 non-QPM lines used in the panel, 19.9% had values higher than 30 μg g^{-1} DW, and about 6% of them had values higher than the breeding target (33 μg g^{-1} DW). Taken together, these results indicate a great potential to develop high-Zn maize varieties either alone or in combination with better protein quality in biofortification programs.

Is there a Trade-Off between Kernel Zn and Grain Yield?

It is generally perceived that nutritional enhancement in crops may result in a yield penalty. QPM, provitamin A, and Zn-enriched maize are no exception to this perception. A lack of correlation between kernel Zn concentration and grain yield was reported in several studies, suggesting that improvement of kernel Zn is possible without reducing grain yield (Bänziger & Long, 2000; Chakraborti et al., 2009a, b). In contrast, a negative correlation between grain yield and kernel Zn was reported by Baxter et al. (2013). The negative correlation may be due to the "dilution effect," whereby in the high-yielding genotypes, increased carbohydrate content in the grain dilutes the concentration of Zn (Bänziger & Long, 2000). However, different studies have demonstrated that both grain yield and nutritional quality traits could be improved simultaneously through breeding (Vivek et al., 2008; Gupta et al., 2013; Muthusamy et al., 2014; Goredema-Matongera et al., 2023).

In a recent study, Goredema-Matongera et al. (2023) compared Zn-enhanced hybrids with different nutritional backgrounds with non-biofortified checks. The highest grain yield observed in non-biofortified commercial checks concurred with other studies reporting dilution effects (Bänziger & Long, 2000; Menkir, 2008). However, the differences in grain yield performance between Zn-enhanced hybrids and commercial checks were not large; this suggests the possibility of developing high-yielding and nutritionally superior maize genotypes. One must not compare the genetic gains through conventional maize breeding with high-Zn maize breeding, as the number of years of breeding, levels of investment, and breeding networks involved in continuous improvement are hugely different. Nevertheless, through well-implemented breeding programs, genotypes with moderate to high yield potential coupled with high levels of grain micronutrients can be developed. Application of foliar Zn-containing fertilizers can also further increase the grain Zn concentration in the case of high-yielding cultivars with a moderate ability to accumulate grain Zn in soil (Liu et al., 2020).

Empirical evidence accumulated in the CIMMYT's Zn enhancement breeding program targeted to the lowland tropics of Latin America shows relative yield parity and similar performance for other agronomic traits of Zn-enhanced hybrids relative to non-Zn commercial checks, indicating the competitiveness of these products in the lowland tropics of Latin America (Table 2).

Molecular Breeding for Kernel Zn Enrichment

Wallace et al. (2014), after an extensive review of many trait mapping studies in maize using genome-wide association studies (GWAS), observed that the majority of the traits in maize are quantitatively inherited, with several small-effect genetic loci with epistatic and environmental interactions controlling them. Based on various studies, it is well established that kernel-Zn concentration is a complex trait, with varying levels of heritability reported in different genetic mapping experiments and significant genetic background and environmental influences. Moreover, there could be many other possibilities for the introduction of variation for the trait in genetic studies, like the quantification methods used, improper postharvest handling, and crop management practices (Garcia-Oliveira et al., 2018). QTL mapping and GWAS are powerful and complementary strategies for developing trait markers for use in marker-based selection. Over the past few years, a few loci that are responsible for kernel-Zn concentration have been detected through QTL mapping in maize. Qin et al. (2012) reported three stable QTLs for kernel-Zn concentrations in two populations across two environments. Similarly, Šimić et al. (2012) identified two minor QTLs on Chr 3.05 and 4.08 in a temperate biparental population phenotyped over two locations. Different QTL mapping studies on kernel-Zn concentration in maize have identified 3–17 QTLs responsible for the trait (Zhou et al., 2010; Lung'aho et al., 2011; Šimić et al., 2012; Qin et al., 2012; Jin et al., 2013; Hindu et al., 2018). A meta-QTL study conducted by Jin et al. (2013), based on five QTL mapping studies published for kernel Zn and many related minerals, identified nine

Table 2 Means for yield and some other agronomic traits across elite lowland tropical CIMMYT high Zn hybrids, highest-yielding Zn and non-Zn hybrids, and commercial checks in advanced multilocation yield trials conducted in Latin America in 2019–2021

	Mean grain yield (t/ha)	Days to anthesis	Ear rot (%)	Root lodging (%)
White hybrids evaluated in 2019 over 20 sites in Mexico, Central America and Colombia				
CIMMYT Zn hybrids (n = 15)	5.79	58.0	10.5	5.7
CIMMYT non-Zn hybrids (n = 1)	6.11	57.1	11.2	5.3
Highest-yielding CIMMYT Zn hybrid	6.02	58.3	8.3	4.1
Highest-yielding CIMMYT non-Zn hybrid	6.11	57.1	11.2	5.3
Commercial Zn Check	5.82	57.1	10.0	2.2
Commercial non-Zn Check1	5.76	57.4	11.4	4.1
Commercial non-Zn Check2	5.65	57.6	11.2	7.4
LSD (0.05)	0.32	0.66	2.91	3.97
Heritability	0.69	0.92	0.64	0.77
White hybrids evaluated in 2020 over 12 sites in Mexico, Central America, and Colombia				
CIMMYT Zn hybrids (n = 14)	4.64	55.6	14.1	4.6
CIMMYT non-Zn hybrids (n = 1)	4.98	55.1	15.9	4.3
Highest-yielding CIMMYT Zn hybrid	4.96	56.2	11.9	4.8
Highest-yielding CIMMYT non-Zn hybrid	4.98	55.1	15.9	4.3
Commercial Zn Check	4.44	55.3	13.9	3.3
Commercial non-Zn Check1	4.91	56.4	14.8	2.7
Commercial non-Zn Check2	4.37	55.4	16.9	4.5
LSD (0.05)	0.46	0.68	4.01	2.97
Heritability	0.71	0.81	0.47	0.41
White hybrids evaluated in 2021 over 18 sites in Mexico, Central America and Colombia				
CIMMYT Zn hybrids (n = 15)	5.24	56.2	7.6	4.8
CIMMYT non-Zn hybrids (n = 2)	5.73	55.3	7.4	4.5
Highest-yielding CIMMYT Zn hybrid	5.57	55.7	7.3	4.8
Highest-yielding CIMMYT non-Zn hybrid	5.90	54.7	7.1	4.7
Commercial Zn Check	5.26	55.7	8.2	5.0
Commercial non-Zn Check1	5.13	56.2	8.4	4.0
Commercial non-Zn Check2	5.28	55.5	7.2	4.1
LSD (0.05)	0.34	0.62	2.07	2.05
Heritability	0.82	0.92	0.71	0.22

meta-QTLs across the maize chromosomes with potential influence on kernel Zn concentration.

Two large GWAS studies have been reported for kernel Zn content in maize. A large association mapping panel of 923 maize inbred lines assembled at CIMMYT, consisting of 432 tropical, 402 subtropical, and 89 temperate germplasm, was evaluated across three environments. The study showed a range of kernel-Zn values from 17.11 to 43.69 µg/g. The panel was genotyped using high-density Genotyping by Sequencing (GBS), identifying a total of 20 SNPs that are significantly associated with kernel-Zn (Hindu et al., 2018). As part of this study, QTL mapping was conducted in three populations to validate the association signals, apart from identifying new QTLs. Eleven of the SNPs identified were validated in one or more of the mapping populations. Several reported QTL studies on kernel-Zn identified QTLs in chromosomal bins 3.04 (Qin et al., 2012), 4.06, 5.04 (Jin et al., 2013), and 9.06–9.07 (Qin et al., 2012; Jin et al., 2013), which were found to have significant SNPs for kernel-Zn identified in CIMMYT GWAS and validation studies. For translating these findings into the CIMMYT high-Zn breeding pipeline, the validated genomic regions and the identified QTLs were further studied in the breeding parents for their allele effects and frequency. For these identified markers, the favorable allele frequency among parents in the breeding pipeline ranged from 0.09 to 0.94. Three haplotypes were selected on chromosomes 5, 7 and 9 based on their favorable allele frequency and effect size of favorable alleles. Selection for these three haplotypes could result in a 16.3% improvement over the population mean. To implement MAS, haplotype optimization and more validation studies in diverse breeding populations involving QPM and non-QPM parents are being carried out.

A more recent GWAS study conducted in a subset of the predominantly temperate North Central Regional Plant Introduction Station association panel (Ames panel), evaluated 2165 maize lines for kernel Zn content in two seasons (Wu et al., 2021). Out of this, a small proportion (31) were also CIMMYT Maize Lines (CMLs), which are of tropical/subtropical origin. The kernel Zn concentrations ranged from 12.5 to 53.4 µg/g, with CML372 showing 42.9 µg/g. The study identified two significant SNPs in moderately strong linkage disequilibrium on chromosome 7, about 1.9Kb from the nicotianamine synthase5 (*nas5*) gene (Zm00001d022557) that codes for a class II NAS reported to be involved in synthesizing the metal ion chelator nicotianamine (Zhou et al., 2013). The identified SNPs are physically located around 500Kb from the haplotype on chromosome 7, which was validated in the CIMMYT study. It is interesting to note that the identified haplotypes in the two studies were also found to have significant association with kernel-Fe, which has been reported to be highly correlated traits (Hindu et al., 2018; Wu et al., 2021). Considering the importance of this gene, gene-based SNP assays are being developed to optimize the favorable haplotypes to select for kernel-Zn in the high-Zn breeding pipeline of CIMMYT. Apart from this, the Ames panel GWAS identified significant association of kernel-Zn with SNPs located 1.2Kb from the *Yellow Stripe Like2* (*ysl2*) gene (Zm00001d017427) on chromosome 5. The protein encoded by its orthologous genes in *Arabidopsis* is shown to transport metal-nicotianamine complexes to various Arabidopsis plant tissues (Waters et al., 2006). The YSL family of transporters uptake metal-phytosiderophores or metalnicotianamine complexes, and Zang et al. (2020) showed that *ZmYSL2* is a metal-nicotianamine transporter involved in the transport of Fe from the endosperm to embryo in the developing maize grain, and in the transport of Zn. Though this genomic region was not detected in the CIMMYT tropical maize GWAS, efforts are ongoing to develop gene-based markers for this gene also to study the allele effect and allele frequency in CIMMYT high-Zn breeding pipeline.

Genomic selection (GS), also known as genomic prediction (GP), is a form of marker-assisted selection in which genetic markers covering the whole genome are used to predict the breeding values of offspring in a population by associating their traits with their high-density genetic marker scores. In GS, a training set, for which phenotypic and genotypic data was generated, is used to estimate the effect of genetic markers covering the whole genome. The marker effects estimated from the training set are then used to predict the genomic estimated breeding value (GEBV) of individuals in the prediction set, which have been genotyped but not phenotyped (Meuwissen et al., 2001). GS/GP has been demonstrated to be an effective approach for improving complex traits that are quantitatively inherited with several small-effect genetic loci, where the effect size of individual locus is not sufficient to improve the target trait by implementing MAS (Prasanna et al., 2021). GS has the potential to reduce phenotyping cost and save breeding time for improving kernel-Zn concentration in the breeding programs, due to the constraints of throughput and cost of phenotyping a large number of individuals in conventional breeding (Prasanna et al., 2020).

The potential of GS for improving kernel-Zn concentration in maize has been investigated in a few studies (Guo et al., 2020; Mageto et al., 2020a, b). In different types of maize populations developed by CIMMYT, moderate to high prediction accuracies for kernel-Zn concentration were observed across different environments, genotyping platforms, cross-validation schemes, prediction models, etc. In the GWAS panel of 923 maize-inbred lines and the two QTL mapping populations (mentioned above), the estimated genomic prediction accuracies for kernel-Zn concentration were 0.69, 0.70, and 0.50, respectively. By incorporating the genotype-by-environment interactions into the prediction model, the prediction accuracy was slightly improved (Mageto et al., 2020a, b). A subset of the GWAS panel of 300 inbred lines and the two QTL mapping populations were also genotyped with repeat amplification sequencing (rAmpSeq) markers, a simple, robust, and cost-effective genotyping platform developed for large-scale GS projects, approximately 8000 markers can be genotyped for US$ 5 per sample. In the same population, the prediction accuracy for kernel-Zn concentration estimated with the rAmpSeq markers was slightly lower than that estimated with the GBS markers, and the prediction accuracy for kernel-Zn concentration estimated with 3000 markers was almost the same as the maximum prediction accuracy estimated with the whole marker dataset (Guo et al., 2020). A multiple-years' training population dataset was built to implement GS for improving grain yield and kernel-Zn concentration simultaneously; the prediction accuracies for kernel-Zn concentration across years ranged from 0.06 to 0.52 (CIMMYT, "unpublished"). These results suggest that GS has the potential to accelerate breeding for improving kernel-Zn concentration; however, the prediction accuracies across years and populations must be assessed in a larger breeding dataset with a closer relationship between the training and prediction sets in further studies.

The three haplotypes identified on chromosomes 5, 7, and 9 (Hindu et al., 2018) were used to predict the kernel-Zn content of lines entering Stage 1 testing, where they showed a moderate prediction accuracy of 0.43 in a five-fold cross-validation study. Fitting these three haplotypes as covariates in the genomic prediction using genome-wide SNPs was not found to improve the prediction accuracy of 0.59 obtained using genome-wide markers alone. Kernel-Zn concentration is a difficult trait to assay in a large number of individuals in a breeding pipeline, and the value of trait markers to enrich the favorable allele frequency in breeding populations has been demonstrated by the shift in population mean of kernel-Zn due to trait marker-based selection. Hence, it is suggested to use the two strategies of haplotype-based selection and genomic selection in a stepwise manner for the improvement of the trait in the breeding pipeline, as follows: (i) At the early stage of the kernel Zn pipeline in F2 or doubled haploids (DH), MAS could be utilize d to enrich the favorable haplotypes on chromosomes 5, 7, and 9 for kernel Zn content; (ii) during Stage 1 testing, GS could be utilize d to improve grain yield and kernel Zn simultaneously by selection of the untested Stage 1 breeding lines using GEBVs. The same genotyping dataset for grain yield prediction will be used for Zn prediction; (iii) rapid cycling at an early stage of the selected lines, which helps enrich the favorable alleles within the breeding population at a faster pace.

Bioavailability of Kernel Zn and Nutritional Impact

Kernel Zn concentrations in maize may show significant variations across the ear development stages. High-zinc kernel genotypes display increased zinc and iron content during the milky stage, when maize is fresh and green. Given the popularity of the consumption of fresh maize, mainly boiled, toasted, or roasted, fresh maize biofortified with zinc can meet up to 89% and 100% of Estimated Average Requirements (EAR) for pregnant women and children, respectively (Rosales et al., 2023).

The nutritional benefits of Zn in maize can be compromised due to the phytate levels in maize itself or in other components of the diet and by the chosen method of maize processing and consumption. This is especially the case in refined maize meal production. Removal of the pericarp and embryo can result in Zn loss. High-Zn maize typically contains up to 33 µg g^{-1} of Zn, compared with 19.5–22.6 µg g^{-1} in conventional maize (Gallego-Castillo et al., 2021; Rosales et al., 2023). The distribution of Zn in grain components, notably the endosperm and embryo, is crucial; roughly 20–30% of the total zinc resides in the endosperm, while about 65–75% is concentrated in the embryo. Preserving the embryo during processing is essential to retaining higher zinc levels in final maize products. Techniques such as alkaline cooking (nixtamalization), widely used in Mesoamerica, utilize whole maize kernels, contributing to greater Zn retention, especially in products like tortillas. This stands in contrast to other maize dishes like arepas or mazamorra, common in countries like Panama, Venezuela, and Colombia (Gallego-Castillo et al., 2021). This information highlights the potential for optimising processing methods to maximize zinc retention and enhance maize-based foods' nutritional value, contributing to improved public health and nutrition.

In addition to optimizing processing techniques, it is essential to consider soil management practices, including crop rotations and diversification, and preserve the nutrient content of maize, particularly the Zn-rich embryo (Gallego-Castillo et al., 2021). Combining high-Zn maize with other crops and legumes can also be a powerful strategy to harness the full nutritional potential of maize and contribute to improved public health and nutrition in regions where maize is a dietary staple.

The study conducted by Chomba et al. (2015) demonstrated that high-Zn maize effectively met the dietary Zn requirements of young rural Zambian children. While research over the past two decades has substantiated the nutritional efficacy and health impact of biofortification for various crops and nutrients, including Zn, further research is needed to evaluate the nutritional value and potential health impacts of high-Zn maize in diverse age groups and contexts. Exploring the synergistic benefits of combining different biofortified crops or integrating them with other nutrient-rich food sources, often referred to as "food-based strategies," holds the potential to address micronutrient deficiencies more comprehensively and sustainably. For instance, combining biofortified maize with legumes or vegetables rich in complementary nutrients can offer further solutions to nutritional challenges. This type of research can lead to innovative dietary interventions that optimize nutrient bioavailability, contributing to improved public health and nutrition outcomes.

High-Zn Maize Varietal Releases and Extent of Adoption

Tropical maize inbred lines with enhanced Zn content and other desirable agronomic and adaptive traits developed at CIMMYT have been used in several countries to develop agronomically competitive hybrids and OPVs (Prasanna et al., 2020). Extensive multi-location evaluations of Zn-enriched OPVs and hybrids in collaboration with public and private sector partners in Mexico, Guatemala, Nicaragua, El Salvador, Honduras, and Colombia led to the selection of promising hybrids and synthetics for further evaluation in national performance trials (NPTs) and farmer participatory on-farm trials. Results showed relative yield parity and similar performance for other agronomic traits relative to commercial checks, indicating the competitiveness of these products in the lowland tropics of Latin America. So far, ten high-Zn maize cultivars (four hybrids and four synthetics) have been released in five countries, including Guatemala, Honduras, Nicaragua, Colombia, and El Salvador (Table 3; Fig. 1). These cultivars have 90–110% of the target kernel Zn content set under HarvestPlus and are competitive with the commercial checks for grain yield and other adaptive traits (Listman et al., 2019). In addition, the high-Zn inbred lines developed at CIMMYT are being used as trait donors by national maize breeding programs and NGOs in Asia (e.g., Nepal, Pakistan, and India) and Latin America (e.g., Colombia and Guatemala) to develop improved, nutritionally

Table 3 Details of high-Zinc maize cultivars released globally (as of August 2023)

S. No.	Variety	OPV/Hybrid?	Country	Released by Partner Institution	Year of Varietal Release
1	DICTA B02	OPV	Honduras	DICTA (NARS)	2017
2	DICTA B03	OPV	Honduras	DICTA (NARS)	2018
3	ICTA HB-18	Hybrid	Guatemala	ICTA (NARS)	2018
4	ICTA B-15	OPV	Guatemala	ICTA (NARS)	2018
5	Fortaleza F5	Hybrid	Guatemala	Semilla Nueva (NGO)	2021
6	Fortaleza F7	Hybrid	Guatemala	Semilla Nueva (NGO)	2022
7	BIOMZN-01	OPV	Colombia	Maxi Semillas (SME)	2018
8	SGBIOH2	Hybrid	Colombia	Semillas Guerrero (SME)	2019
9	Centa Porrillo 2020	OPV	El Salvador	CENTA (NARS)	2020
10	Fortinica	OPV	Nicaragua	INTA (NARS)	2018

Note: All the ten high-Zn maize cultivars released in Latin America are in QPM background

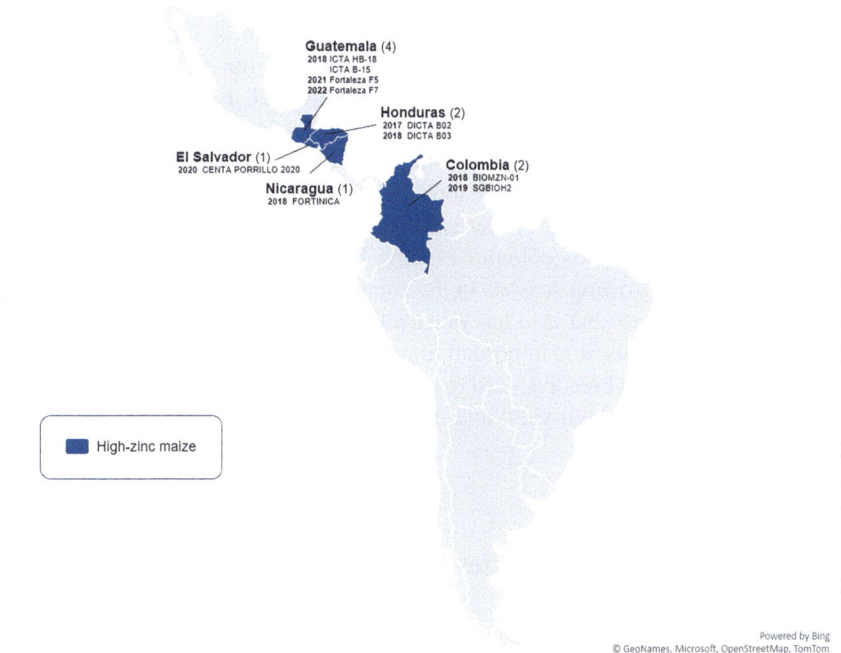

Fig. 1 Map showing countries with high-Zn maize cultivars released for commercial cultivation in Latin America

enriched maize cultivars (Maqbool & Beshir, 2019; Listman et al., 2019). Moreover, tropical high-Zn donors are being used by CIMMYT's Asia and SSA regional maize breeding programs for developing high-Zn cultivars.

In Guatemala, Zn-enriched maize varieties ICTA HB-18 and ICTA B-15 were developed by CIMMYT as a part of the CGIAR Research Program on Maize (MAIZE) and the CGIAR Research Program on Agriculture for Nutrition and Health (A4NH)-HarvestPlus, in partnership with Guatemala's Institute for Agricultural Science and Technology (ICTA). Guatemala is the first country to release a Zn-enriched maize hybrid. These cultivars were developed through conventional breeding. Farmers are highly attracted to these varieties due to their yield potential, high-quality protein content, high Zn concentration, early maturity, and large kernel size. ICTA HB-18 has 15% more kernel Zn than other commercial varieties, whereas tortillas made with ICTA B-15 have almost 60% more Zn than tortillas made from other commercial varieties (Johnson, 2018a). More recently, Semilla Nueva, a Guatemalan-based NGO, released two three-way cross hybrids (Fortaleza 5 and Fortaleza 7) using CIMMYT's high-Zn inbred lines. These new hybrids have better yield potential and kernel Zn concentration parity with the reference commercial check (ICTA HB-18). Approximately 9% (12,000 bags/240 MT) of all white maize hybrid seeds sold in Guatemala in 2023 will be biofortified with high-Zn (Curt Bowen, "personal communication"). This success story has been possible due to a subsidy program for seed purchases implemented by Semilla Nueva.

BIO-MZN01 is a Zn-enriched maize variety developed by CIMMYT with the collaboration of HarvestPlus, the International Center for Tropical Agriculture (CIAT), CGIAR Research Programs MAIZE and A4NH. BIO-MZN01 has 36% more kernel Zn than other varieties, meaning that arepas, one of the Colombian dishes, may have up to five times more Zn when made from this new variety. BIO-MZN01 has yield potential of up to 8 tons per hectare (t/ha) and is tolerant to major maize diseases such as rust, grey leaf spot, and Turcicum leaf blight. It can also be grown across diverse agro-ecologies ranging from 0 to 1400 m above sea level (masl) during both growing seasons in the country (Johnson, 2018b). BIO-MZN01 not only is Zn-enriched but also has other relevant traits, including high yield and disease resistance. Thus, it is important to note that only kernel Zn enrichment will not lead to success, but breeders need to ensure that the variety has a desirable package of traits, including high yield and disease resistance, to compete in the target market segment.

Challenges and Opportunities

In the twenty-first century, the challenge is not only to produce enough quantity of food to feed the growing population but also to provide nutritionally balanced diets while respecting the cultural backgrounds of consumers. The food on everyone's plates must be of appropriate quantity, nutritious, and produced in an environmentally, economically, and socially sustainable manner. The EAT-Lancet Commission Report (Willet et al., 2019) highlighted the importance of promoting diets that are nutritious and that can reduce the environmental impact of food systems.

Although not a silver bullet solution, biofortification is an effective strategy to combat micronutrient malnutrition and can be complementary to other approaches, including fortification, crop and dietary diversification. CIMMYT, IITA, and national partners (especially in Africa, Asia, and Latin America) have employed conventional breeding and molecular tools to successfully breed and release several nutritionally enriched maize cultivars over the years without compromising on grain yield levels or other important agronomic and adaptive traits. The rapid advances that have been made in understanding the genetic control of zinc concentration in maize kernels, coupled with the availability of new tools and technologies (e.g., genomic selection and prediction; high-throughput phenotyping), could potentially accelerate the rate of genetic gain for improved kernel Zn content in maize breeding programs.

Several challenges remain, as highlighted below, for breeding and deploying high-Zn maize in the low- and middle-income countries in the Global South:

- Breeding for biofortified crop varieties is indeed gender-responsive, as enriching the micronutrient contents of staple foods can have a huge impact on the health and well-being of women and children. However, high-Zinc maize cannot be distinguished in the field from normal maize with low-Zn content, unlike disease resistance, drought/heat tolerance, or insect-pest resistance. This "invisibility" of the trait poses a challenge, especially when promoting high-Zn maize cultivars to farming communities. The benefits of high-Zn maize in terms of nutritional well-being will only be visible after adoption and consumption over time. Moreover, unlike an "orange maize with high provitamin A content," it is not possible to "brand" a high-Zinc maize using visible traits.
- Breeding for high-Zn maize cannot succeed without incorporating other important traits demanded by the farming communities, including high yield, yield stability, resistance to major diseases, and tolerance to climate-induced stresses (e.g., drought, heat, etc.), depending on the target market segment. Breeding for high-Zn maize must also take into consideration important traits demanded by consumers, including grain texture and size. Unfortunately, investment in breeding pipelines that target nutritional enrichment in maize (e.g., provitamin A, kernel Zn) has significantly declined over the years, limiting progress through biofortification.
- Research is required on the analysis of the nutritional impact of integrating different knowledge systems, such as multi-crop production systems, utilisation of biofortified varieties, and culturally accepted food processing methods that preserve micronutrients (e.g., by utilising whole grains).
- Some of the high-Zn maize cultivars are currently grown by farmers in Latin America (albeit to a limited extent), while limited efforts are underway to develop and deploy high-Zn maize cultivars with farmer-preferred traits in SSA and Asia. However, more intensive integration of national and international research efforts and interdisciplinary work is key to the wider adoption of high-Zinc maize cultivars with a significant nutritional impact on consumers.

- It is important to understand the maize value chains and market dynamics when deploying nutritionally enriched cultivars, such as high-Zn maize. Maize value chains that effectively link the farmers who grow biofortified varieties to the processors on the one hand and government programs focusing on alleviating malnutrition on the other hand are indeed vital. Only through such linkages can the value of biofortified cultivars be fully exploited, malnutrition alleviated, and new markets opened.
- Intensive efforts are needed to create awareness among policymakers and the public about the availability of biofortified crop varieties and their potential linkages with the nutritional education programs of governments.
- Resource mobilization is key for building strong breeding pipelines for biofortification, for generating impactful products, and for sustainable adoption of biofortified varieties, such as high-Zn maize, by the farming communities in targeted countries in the Global South.

Conclusions

Micronutrient malnutrition, including zinc deficiency, is a silent epidemic affecting a significant proportion of the world's population and has a particularly high negative impact on populations in low- and middle-income countries. Biofortification for enhanced kernel zinc content in maize is possible, as demonstrated by CGIAR Centers and national partners in the last 10–15 years, through systematic screening of germplasm, genetic dissection of the target trait, and the development of high-Zn inbred lines and cultivars. At the same time, investments in biofortification programs across crops, including maize, have been suboptimal. This can have huge negative consequences unless nutritional enrichment through breeding or biofortification is sufficiently funded to create sustainable impacts.

The success stories of other biofortified crops, including high-Zn wheat, high-Fe beans, and high-provitamin A ("orange flesh") sweet potatoes, and the emerging success story of high-Zn maize from Guatemala, reflect the importance of public-private partnerships in stimulating demand for biofortified varieties. This needs to be further scaled up and replicated in several target countries across Latin America, sub-Saharan Africa, and Asia through multi-institutional synergies. There must also be a greater recognition of the role of nutritional education and that biofortification could be an effective component of a multipronged strategy for improving the nutritional well-being of populations globally.

References

Agedew, E., Tsegaye, B., Bante, B., Zerihun, E., Aklilu, A., Girma, M., Kerebih, H., Wale, M. Z., & Yirsaw, M. T. (2022). Zinc deficiency and associated factors among pregnant women's attending antenatal clinics in public health facilities of Konso Zone, Southern Ethiopia. *PLoS One, 17*(7), e0270971. https://doi.org/10.1371/journal.pone.0270971

Agrawal, P. K., Jaiswal, S. K., Prasanna, B. M., Hossain, F., Saha, S., Guleria, S. K., & Gupta, H. S. (2012). Genetic variability and stability for kernel iron and zinc concentration in maize (*Zea mays* L.) genotypes. *Indian Journal of Genetics, 72*, 421–428.

Ahsan, A. K., Tebha, S. S., Sangi, R., Kamran, A., Zaidi, Z. A., Haque, T., & Hamza, M. S. A. (2021). Zinc micronutrient deficiency and its prevalence in malnourished pediatric children as compared to well-nourished children: A nutritional emergency. *Global Pediatric Health, 8*, 2333794X211050316. https://doi.org/10.1177/2333794X211050316

Akhtar, S., Osthoff, G., Mashingaidze, K., & Labuschagne, M. (2018). Iron and zinc in maize in the developing world: Deficiency, availability, and breeding. *Crop Science, 58*, 2200–2213.

Akinwale, R. O., & Adewopo, O. A. (2016). Grain iron and zinc concentrations and their relationship with selected agronomic traits in early and extra-early maize. *Journal of Crop Improvement, 30*, 641–656.

Arnold, J. M., & Bauman, L. F. (1976). Inheritance of and interrelationships among maize kernel traits and elemental contents. *Crop Science, 16*, 439–440.

Arnold, J. M., Bauman, L. F., & Aycock, H. S. (1977). Interrelations among protein, lysine, oil, certain mineral element concentrations, and physical kernel characteristics in two maize populations. *Crop Science, 17*, 421–425. https://doi.org/10.2135/cropsci1977.0011183X001700030020x

Bänziger, M., & Long, J. (2000). The potential for increasing the iron and zinc density of maize through plant-breeding. *Food and Nutrition Bulletin, 21*, 397–400.

Baxter, I. R., Gustin, J. L., Settles, A. M., & Hoekenga, O. A. (2013). Ionomic characterization of maize kernels in the intermated B73 x Mo17 population. *Crop Science, 53*, 208–220.

Birol, E., & Bouis, H. E. (2023). Role of socio-economic research in developing, delivering and scaling new crop varieties: The case of staple crop biofortification. *Frontiers in Plant Science, 14*, 1099496.

Bouis, H. E., & Saltzman, A. (2017). Improving nutrition through biofortification: A review of evidence from HarvestPlus, 2003 through 2016. *Global Food Security, 12*, 49–58.

Bouis, H. E., Hotz, C. H., McClafferty, B., Meenakshi, J. V., & Pfeiffer, W. (2011). Biofortification: A new tool to reduce micronutrient malnutrition. *Food and Nutrition Bulletin, 32*, S31–S40.

Brkic, I., Simic, D., Zdunic, Z., Jambrovic, A., Ledencan, T., Kovacevic, V., & Kadar, I. (2004). Genotypic variability of micronutrient element concentrations in maize kernels. *Cereal Research Communications, 32*, 107–112.

Brown, K. H., Rivera, J. A., Bhutta, Z., Gibson, R. S., King, J. C., Lonnerdal, B., Ruel, M. T., Sandtrom, B., Wasantwisut, E., & Hotz, C. (2004). International Zinc Nutrition Consultative Group (IZiNCG) technical document #1. Assessment of the risk of zinc deficiency in populations and options for its control. *Food and Nutrition Bulletin, 25*, 91–204.

Cakmak, I. (2008). Enrichment of cereal grains with zinc: Agronomic or genetic biofortification? *Plant and Soil, 302*, 1–17.

Caulfield, L. E., & Black, R. E. (2004). Zinc deficiency. In M. Ezzati, A. D. Lopez, C. J. Murray, & A. Rodgers (Eds.), *Comparative quantification of health risks: Global and regional burden of diseases attributable to selected major risks* (pp. 257–279). World Health Organization.

Chakraborti, M., Hossain, F., Kumar, R., Gupta, H. S., & Prasanna, B. M. (2009a). Genetic evaluation of grain yield and kernel micronutrient traits in maize. *Pusa AgriScience, 32*, 11–16.

Chakraborti, M., Prasanna, B. M., Hossain, F., Singh, A. M., & Guleria, S. K. (2009b). Genetic evaluation of kernel Fe and Zn concentrations and yield performance of selected maize (*Zea mays* L.) genotypes. *Range Management and Agroforestry, 30*, 109–114.

Chakraborti, M., Prasanna, B. M., Hossain, F., Mazumdar, S., Singh, A. M., Guleria, S., & Gupta, H. S. (2011a). Identification of kernel iron- and zinc-rich maize inbreds and analysis of genetic diversity using microsatellite markers. *Journal of Plant Biochemistry and Biotechnology, 20*, 224–233.

Chakraborti, M., Prasanna, B. M., Hossain, F., & Singh, A. M. (2011b). Evaluation of single cross quality protein maize (QPM) hybrids for kernel iron and zinc concentrations. *Indian Journal of Genetics and Plant Breeding, 71*, 312–319.

Chomba, E., Westcott, C. M., Westcott, J. E., Mpabalwani, E. M., Krebs, N. F., Patinkin, Z. W., Palacios, N., & Hambidge, K. M. (2015). Zinc absorption from biofortified maize meets the requirements of young rural Zambian children. *The Journal of Nutrition, 145*, 514–519.

CIMMYT. (1998). *A complete listing of improved maize germplasm from CIMMYT. Maize program special report*. CIMMYT.

Council for Agricultural Science and Technology (CAST). (2020). *Food biofortification—Reaping the benefits of science to overcome hidden hunger—A paper in the series on the need for agricultural innovation to sustainably feed the world by 2050. Issue paper 69*. CAST.

de Benoist, B., Darnton-Hill, I., Davidsson, L., Fontaine, O., & Hotz, C. (2007). Conclusions of the joint WHO/UNICEF/IAEA/IZiNCG interagency meeting on zinc status indicators. *Food and Nutrition Bulletin, 28*, S480–S484.

de Valença, A. W., Bake, A., Brouwer, I. D., & Giller, K. E. (2017). Agronomic biofortification of crops to fight hidden hunger in sub-Saharan Africa. *Global Food Security, 12*, 8–14.

Diez-Altares, C., & Bornemisza, E. (1967). The localization of zinc-65 in germinating corn tissues. *Plant and Soil, 26*, 175–188.

Ekpa, O., Palacios-Rojas, N., Kruseman, G., Fogliano, V., & Linnemann, A. (2019). Sub-saharan African maize-based foods: Processing practices, challenges and opportunities. *Food Review International*. https://doi.org/10.1080/87559129.2019.1588290

Gallego-Castillo, S., Taleon, V., Talsma, E. F., Rosales-Nolasco, A., & Palacios-Rojas, N. (2021). Effect of maize processing methods on the retention of minerals, phytic acid and amino acids when using high kernel-zinc maize. *Current Research in Food Science, 4*, 279–286.

Garcia-Oliveira, A. L., Chander, S., Ortiz, R., Menkir, A., & Gedil, M. (2018). Genetic basis and breeding perspectives of grain iron and zinc enrichment in cereals. *Frontiers in Plant Science, 9*, 937.

Goredema-Matongera, N., Ndhlela, T., van Biljon, A., Kamutando, C. N., Cairns, J. E., Baudron, F., & Labuschagne, M. (2023). Genetic variation of zinc and iron concentration in normal, provitamin A and quality protein maize under stress and non-stress conditions. *Plants, 12*, 270. https://doi.org/10.3390/plants12020270

Gorsline, G. W., Thomas, W. I., & Baker, D. E. (1964). Inheritance of P, K, Mg, Cu, B, Zn, Mn, Al, and Fe concentrations by corn (*Zea mays* L.) leaves and grain. *Crop Science, 4*, 207–210.

Guleria, S. K., Chahota, R. K., Kumar, P., Kumar, A., Prasanna, B. M., Hossain, F., & Gupta, H. S. (2013). Analysis of genetic variability and genotype × year interactions on kernel zinc concentration in selected Indian and exotic maize (*Zea mays* L.) genotypes. *Indian Journal of Agricultural Sciences, 83*, 836–841.

Guo, R., Dhliwayo, T., Mageto, E., Palacios-Rojas, N., Lee, M., Yu, D., Ruan, Y., Zhang, A., San Vicente, F., Olsen, M., Crossa, J., Boddupalli, P., Zhang, L., & Zhang, X. (2020). Genomic prediction of kernel zinc content in multiple maize populations using genotyping-by-sequencing and repeat amplification sequencing markers. *Frontiers in Plant Science, 11*, 534.

Gupta, H. S., Raman, B., Agrawal, P. K., Mahajan, V., Hossain, F., & Nepolean, T. (2013). Accelerated development of quality protein maize hybrid through marker-assisted introgression of opaque-2 allele. *Plant Breeding, 132*, 77–82.

Guzmán, C., Medina-Larqué, A. S., Velu, G., González-Santoyo, H., Singh, R. P., Huerta-Espino, J., et al. (2014). Use of wheat genetic resources to develop biofortified wheat with enhanced grain zinc and iron concentrations and desirable processing quality. *Journal of Cereal Science, 60*, 617–622.

Habben, J. E., Kirleis, A. W., & Larkins, B. A. (1993). The origin of lysine-containing proteins in opaque-2 maize endosperm. *Plant Molecular Biology, 23*, 825–838.

Hindu, V., Palacios-Rojas, N., Babu, R., Suwarno, W. B., Rashid, Z., Usha, R., & Nair, S. K. (2018). Identification and validation of genomic regions influencing kernel zinc and iron in maize. *Theoretical and Applied Genetics, 131*, 1443–1457. https://doi.org/10.1007/s00122-018-3089-3

Hummel, M., Hallahan, B. F., Brychkova, G., et al. (2018). Reduction in nutritional quality and growing area suitability of common bean under climate change induced drought stress in Africa. *Scientific Reports, 8*, 16187. https://doi.org/10.1038/s41598-018-33952-4

Jin, T., Zhou, J., Chen, J., et al. (2013). The genetic architecture of zinc and iron content in maize grains as revealed by QTL mapping and meta-analysis. *Breeding Science, 63*, 317–324.

Johnson, J. (2018a). *First zinc-enriched maize in Guatemala to combat malnutrition*. Retrieved from https://www.cimmyt.org/news/first-zinc-enriched-maize-in-guatemala-to-combat-malnutrition/

Johnson, J. (2018b). *First zinc maize variety launched to reduce malnutrition in Colombia*. Retrieved from https://www.cimmyt.org/news/first-zinc-maize-variety-launched-to-reduce-malnutrition-in-colombia/

Krebs, N. F., Miller, L. V., & Hambidge, K. M. (2014). Zinc deficiency in infants and children: A review of its complex and synergistic interactions. *Paediatrics and International Child Health, 34*, 279–288.

Listman, G. M., Guzman, C., Palacios-Rojas, N., Pfeiffer, W., San Vicente, F., & Govidan, V. (2019). Improving nutrition through biofortification: Preharvest and postharvest technologies. *Cereal Food World, 64*, 3.

Liu, D. Y., Zhang, W., Liu, Y. M., Chen, X. P., & Zou, C. Q. (2020). Soil application of zinc fertilizer increases maize yield by enhancing the kernel number and kernel weight of inferior grains. *Frontiers in Plant Science, 11*, 188.

Long, J. K., Banziger, M., & Smith, M. E. (2004). Diallel analysis of grain iron and zinc density in Southern African adapted maize inbreds. *Crop Science, 44*, 2019–2026.

Lopez-Ridaura, S., Barba-Escoto, L., Reyna, C., Sum, C., Palacios-Rojas, N., & Gerard, B. (2021). The milpa system and its contribution to nutrition in the Western Highlands of Guatemala. *Scientific Reports, 11*, 3696. https://doi.org/10.1038/s41598-021-82784-2

Lung'aho, M. G., Mwaniki, A. M., Szalma, S. J., et al. (2011). Genetic and physiological analysis of iron biofortification in maize kernels. *PLoS One, 6*, 1–10.

Mageto, E., Crossa, J., Perez-Rodriguez, P., Dhliwayo, T., Palacios-Rojas, N., Lee, M., Guo, R., San Vicente, F., Zhang, X., & Hindu, V. (2020a). Genomic prediction with genotype by environment interaction analysis for kernel zinc concentration in tropical maize germplasm. *G3*. https://doi.org/10.1101/2020.02.24.963090

Mageto, E., Lee, M., Dhliwayo, T., Palacios-Rojas, N., San Vicente, F., Burgueño, J., & Hallauer, A. R. (2020b). An evaluation of kernel zinc in hybrids of elite quality protein maize (QPM) and non-QPM inbred lines adapted to the tropics based on a mating design. *Agronomy, 10*(5), 695. https://doi.org/10.3390/agronomy10050695

Mallikarjuna, M. G., Nepolean, T., Hossain, F., Manjaiah, K. M., Singh, A. M., & Gupta, H. S. (2014). Genetic variability and correlation of kernel micronutrient among exotic quality protein maize inbreds and their utility in breeding programme. *Indian Journal of Genetics and Plant Breeding, 74*, 166–173.

Maqbool, M. A., & Beshir, A. R. (2019). Zinc biofortification of maize (*Zea mays* L.): Status and challenges. *Plant Breeding, 138*, 1–28. https://doi.org/10.1111/pbr.12658

Marcos-Barbero, E. L., Pérez, P., Martínez-Carrasco, R., Arellano, J. B., & Morcuende, R. (2021). Genotypic variability on grain yield and grain nutritional quality characteristics of wheat grown under elevated CO_2 and high temperature. *Plants, 10*, 1043.

Mayor-Ibarguren, A., & Robles-Marhuenda, A. (2020). A hypothesis for the possible role of zinc in the immunological pathways related to COVID-19 infection. *Frontiers in Immunology, 11*, 1736.

Maziya-Dixon, B., Kling, J. G., Menkir, A., & Dixon, A. (2000). Genetic variation in total carotene, iron, and zinc contents of maize and cassava genotypes. *Food and Nutrition Bulletin, 21*, 419–422.

McDonald, G. K., Genc, Y., & Graham, R. D. (2008). A simple method to evaluate genetic variation in grain zinc concentration by correcting for differences in grain yield. *Plant and Soil, 306*, 49–55. https://doi.org/10.1007/s11104-008-9555-y

Menkir, A. (2008). Genetic variation for grain mineral content in tropical-adapted maize inbred lines. *Food Chemistry, 110*, 454–464.

Meuwissen, T. H. E., Hayes, B. J., & Goddard, M. E. (2001). Prediction of total genetic value using genome-wide dense marker maps. *Genetics, 157*, 1819–1829.

Muthusamy, V., Hossain, F., Nepolean, T., Choudhary, M., Saha, S., Bhat, J. S., Prasanna, B. M., & Gupta, H. S. (2014). Development of β-carotene rich maize hybrids through marker-assisted introgression of β-carotene hydroxylase allele. *PLoS One, 9*(12), e113583.

Myers, S., Zanobetti, A., Kloog, I., et al. (2014). Increasing CO_2 threatens human nutrition. *Nature, 510*, 139–142.

Oikeh, S. O., Menkir, A., Maziya-Dixon, B., Welch, R., & Glahn, R. P. (2003a). Assessment of concentrations of iron and zinc and bioavailable iron in grains of early maturing tropical maize varieties. *Journal of Agricultural and Food Chemistry, 51*, 3688–3694. https://doi.org/10.1021/jf0261708

Oikeh, S. O., Menkir, A., Maziya-Dixon, B., Welch, R., & Glahn, R. P. (2003b). Genotypic differences in concentration and bioavailability of kernel-iron in tropical maize varieties grown under field conditions. *Journal of Plant Nutrition, 26*, 2307–2319. https://doi.org/10.1081/PLN-120024283

Ortiz-Monasterio, J. I., Palacios-Rojas, N., Meng, E., Pixley, K., Trethowan, R., & Peña, R. J. (2007). Enhancing the mineral and vitamin content of wheat and maize through plant breeding. *Journal of Cereal Science, 46*, 293–307. https://doi.org/10.1016/j.jcs.2007.06.005

Owino, V., Kumwenda, C., Ekesa, B., Parker, M. E., Ewoldt, L., Roos, N., Lee, W. T., & Tome, D. (2022). The impact of climate change on food systems, diet quality, nutrition, and health outcomes: A narrative review. *Frontiers in Climate, 4*, 941842.

Palacios-Rojas, N., McCulley, L., Ml, K., Titcomb, T. J., Gunaratna, N. S., Lopez-Ridaura, S., & Tanumihardjo, S. A. (2020). Mining maize diversity and improving its nutritional aspects within agro-food systems. *Comprehensive Reviews in Food Science and Food Safety, 19*, 1809–1834.

Pandey, N., Hossain, F., Kumar, K., Vishwakarma, A. K., Muthusamy, V., Manjaiah, K. M., & Gupta, H. S. (2015). Microsatellite marker-based genetic diversity among quality protein maize (QPM) inbreds differing for kernel iron and zinc. *Molecular Plant Breeding, 6*, 1–10. https://doi.org/10.5376/mpb.2015.06.0003

Pfeiffer, W. H., & McClafferty, B. (2007a). HarvestPlus: Breeding crops for better nutrition. *Crop Science, 50*, S88–S105.

Pfeiffer, W. H., & McClafferty, B. (2007b). Biofortification: Breeding micronutrient-dense crops. In M. S. Kang (Ed.), *Breeding major food staples*. Blackwell Publishing Ltd.

Pixley, K. V., Palacios, N., & Glahn, R. P. (2011). The usefulness of iron bioavailability as a target trait for breeding maize (*Zea mays* L.) with enhanced nutritional value. *Field Crops Research, 123*, 153–160. https://doi.org/10.1016/j.fcr.2011.05.011

Prasanna, B. M., Mazumdar, S., Chakraborti, M., Hossain, F., Manjaiah, K. M., Agrawal, P. K., et al. (2011). Genetic variability and genotype × year interactions for kernel iron and zinc concentration in maize (*Zea mays* L.). *Indian Journal of Agricultural Sciences, 81*, 704–711.

Prasanna, B. M., Palacios-Rojas, N., Firoz, H., Muthusamy, V., Menkir, A., Dhliwayo, T., Ndhela, T., San Vicente, F., Nair, S., Vivek, B., Zhang, X., Olsen, M., & Xingming, F. (2020). Molecular breeding for nutritionally enriched maize: Status and prospects. *Frontiers in Genetics, 10*, 1–16.

Prasanna, B. M., Cairns, J. E., Zaidi, P. H., Beyene, Y., Makumbi, D., Gowda, M., Magorokosho, C., Zaman-Allah, M., Olsen, M., Das, A., Worku, M., Gethi, J., Vivek, B. S., Nair, S. K., Rashid, Z., Vinayan, M. T., Issa, B. A., San Vicente, F., Dhliwayo, T., & Zhang, X. (2021).

Beat the stress: Breeding for climate resilience in maize for the tropical rainfed environments. *Theoretical and Applied Genetics, 134,* 1729–1752.

Qin, H. N., Cai, Y. L., Liu, Z. Z., Wang, G. Q., Wang, J. G., Guo, Y., & Wang, H. (2012). Identification of QTL for zinc and iron concentration in maize kernel and cob. *Euphytica, 187,* 345–358.

Queiroz, V. A. V., Guimaraes, P. E. O., Queiroz, L. R., Guedes, E. O., Vasconcelos, V. D. B., Guimares, L. J., & Schaffert, R. E. (2011). Iron and zinc variability in maize lines. *Ciência e Tecnologia de Alimentos Campinas, 31,* 577–583. https://doi.org/10.1590/S0101-20612011000300005

Rosales, A., Molina-Macedo, A., Leyva, M., San Vicente, F., & Palacios-Rojas, N. (2023). Fresh/high-Zinc maize: A promising solution for alleviating zinc deficiency through significant micronutrient accumulation. *Food, 12*(14), 2757.

Schmidt, R. J., Burr, F. A., Aukerman, M. J., & Burr, B. (1990). Maize regulatory gene opaque-2 encodes a protein with a "leucine-zipper" motif that binds to zein DNA. *Proceedings of the National Academy of Sciences of the United States of America, 87,* 46–50.

Šimić, D., Sudar, R., Ledenčan, T., Jambrović, A., Zdunić, Z., Brkić, I., & Kovačević, V. (2009). Genetic variation of bioavailable iron and zinc in grain of a maize population. *Journal of Cereal Science, 50,* 392–397.

Šimić, D., Drinić, S. M., Zdunić, Z., Jambrović, A., Ledenčan, T., Brkić, J., Brkić, A., & Brkić, I. (2012). Quantitative trait loci for biofortification traits in maize grain. *The Journal of Heredity, 103,* 47–54.

Smith, M. R., & Myers, S. S. (2018). Impact of anthropogenic CO_2 emissions on global human nutrition. *Nature Climate Change, 8,* 834–839. https://doi.org/10.1038/s41558-018-0253-3

Thakur, N., Kumari, R., Prakash, J., Sharma, J., Singh, N., & Lata, S. (2015). Evaluation of elite maize genotypes (*Zea mays* L.) for nutritional traits. *Electronic Journal of Plant Breeding, 6,* 350–354.

Vivek, B. S., Krivanek, A. F., Palacios-Rojas, N., Twumasi-Afriyie, S., & Diallo, A. O. (2008). *Breeding quality protein maize (QPM): Protocols for developing QPM cultivars.* CIMMYT.

Wallace, J. G., Bradbury, P. J., Zhang, N., Gibon, Y., Stitt, M., et al. (2014). Association mapping across numerous traits reveals patterns of functional variation in maize. *PLoS Genetics, 10,* e1004845.

Waters, B. M., Chu, H.-H., Didonato, R. J., Roberts, L. A., Eisley, R. B., et al. (2006). Mutations in Arabidopsis yellow stripe-like1 and yellow stripe-like3 reveal their roles in metal ion homeostasis and loading of metal ions in seeds. *Plant Physiology, 141,* 1446–1458.

Welch, R. M., Smith, M. E., van Campen, D. R., & Schaefer, S. C. (1993). Improving mineral reserves and protein quality of maize (*Zea mays* L.) kernels using unique genes. *Plant and Soil, 156,* 215–218.

Willet, W., Rockstrom, J., Roken, B. et al. (2019). Food in the Anthropocene: the EAT–Lancet Commission onhealthy diets from sustainable food systems. *The Lancet, 393,* 447–492.

Wu, D., Tanaka, R., Li, X., Ramstein, G. P., Cu, S., Hamilton, J. P., Buell, C. R., Stangoulis, J., Rocheford, T., & Gore, M. A. (2021). High resolution genome-wide association study pinpoints metal transporter and chelator genes involved in the genetic control of element levels in maize grain. *G3, 11,* jkab059.

Zang, J., Huo, Y., Liu, J., Zhang, H., Liu, J., et al. (2020). Maize YSL2 is required for iron distribution and development in kernels. *Journal of Experimental Botany, 71,* 5896–5910.

Zhou, J. F., Huang, Y. Q., Liu, Z. Z., Chen, J. T., Zhu, L. Y., Song, Z. Q., & Zhao, Y. F. (2010). Genetic analysis and QTL mapping of zinc, iron, copper and manganese contents in maize seed. *Journal of Plant Genetic Resources, 11,* 593–595.

Zhou, X., Li, S., Zhao, Q., Liu, X., Zhang, S., et al. (2013). Genome-wide identification, classification and expression profiling of nicotianamine synthase (NAS) gene family in maize. *BMC Genomics, 14,* 238.

Open Access This chapter is licensed under the terms of the Creative Commons Attribution 4.0 International License (http://creativecommons.org/licenses/by/4.0/), which permits use, sharing, adaptation, distribution and reproduction in any medium or format, as long as you give appropriate credit to the original author(s) and the source, provide a link to the Creative Commons license and indicate if changes were made.

The images or other third party material in this chapter are included in the chapter's Creative Commons license, unless indicated otherwise in a credit line to the material. If material is not included in the chapter's Creative Commons license and your intended use is not permitted by statutory regulation or exceeds the permitted use, you will need to obtain permission directly from the copyright holder.

Zinc Wheat Variety Release, Seed Production, and Scaling Up Strategies in India

Chandra Nath Mishra, Amit Sharma, Satish Kumar, Disha Kamboj, Gyanendra Singh, Arun Kumar Joshi, and Gyanendra Pratap Singh

Introduction

The value of high-quality seeds has been understood in India since ancient times. In crop production, seed has always remained a fundamental and essential component. According to Manu Smriti, one of the oldest scriptures in India, "Subeejam Sukshetre Jayate Sampadyathe," or "high-quality seed on good soil yields abundantly," describes the significance of seed (Komala et al., 2017). For agrarian communities, seed quality has been revered as a sacred resource that is essential to the development and advancement of agriculture.

According to the Indian Seed Act of 1966, "seed" is legally any of the following classifications used for planting or sowing: (i) seeds of food crops, including edible oil seeds and seeds of fruits and vegetables; (ii) cotton fibers; (iii) seeds of cattle fodder, which also includes seedlings, tubers, bulbs, rhizomes, roots, cuttings, all varieties of grafts, and other asexually propagated material, such as seeds of food crops or cattle fodder (Gadwal, 2003).

The genetic purity of the variety must be maintained throughout the entire seed manufacturing process, and basic guidelines must be followed for greater seed output and purity (Smith & Register III, 1998). These principles for maintenance breeding have been broadly divided into two categories, viz., genetic principles and agronomic seed production principles (Yadav et al., 2022). In this book chapter, we

---

C. N. Mishra (✉) · A. Sharma · S. Kumar · D. Kamboj · G. Singh
ICAR-Indian Institute of Wheat and Barley Research, Karnal, India
e-mail: Chandra.Mishra@icar.gov.in

A. K. Joshi
Borlaug Institute of South Asia, New Delhi, India

G. P. Singh
ICAR-Indian Institute of Wheat and Barley Research, Karnal, India

ICAR-National Bureau of Plant Genetic Resources, New Delhi, India

examine the general needs for transcending wheat seeds, with a focus on wheat biofortification to enrich further for alimentation.

Classes of Seeds

The essential suggestions for high-quality, value-added seed production may improve farmer profitability. Three major types of seeds in India are breeder's, foundation-derived, and certified (Witcombe et al., 1998). The development of a given class of seed from a specified class up to the certified stage is known as the generation system of seed. Preserving genetic purity and maintaining varietal uniqueness are the goals of the generation system of seed multiplication and certification. Farmers purchase the certified or truthfully labelled (TL) seeds to grow the commercial crops. The choice of several generation models is based on factors such as seed need, seed multiplication ratio, and rate of genetic deterioration in a particular variety. Other sorts of seeds in this generation include nucleus and accurately labelled seeds.

Creation of Breeder's and Nucleus Seeds

The original plant breeder or a qualified plant breeder from the institute is in charge of early-stage seed production. From the chosen individual plants of the concerned variety, a small number of seeds are initially harvested, serving as the foundation for the subsequent classes of seed production. The autogamous behaviour of wheat makes seed production reasonably comfortable, although to maintain the highest genetic and physical purity, suggested operational measures should be followed in accordance with the criteria from Harrington (1952) for nucleus seeds (Fig. 1).

I. Variety sampling to acquire nucleus seeds: New, extremely promising lines are sampled for seed purification based on how well they perform in breeding trials or nurseries. Depending on the available resources, a station may only have 15 lines or types of crops in a given year.

II. Table sample examination: Every 200 samples need to be separately threshed and evaluated in stacks on the table, and any that seem unsatisfactory should be thrown away. The nucleus seed, or seed of chosen plant samples of 200 or less, is now prepared for sowing in variety.

III. Nucleus seed location and sowing
If the new variety is released, it should be cultivated in an area that is clean, productive, and suitable for planting the nucleus seed. The location selected for seed production should not have provided the same crop planting for harvesting during the year before.

IV. Examining the nucleus, double-row plots, and eliminating off-type

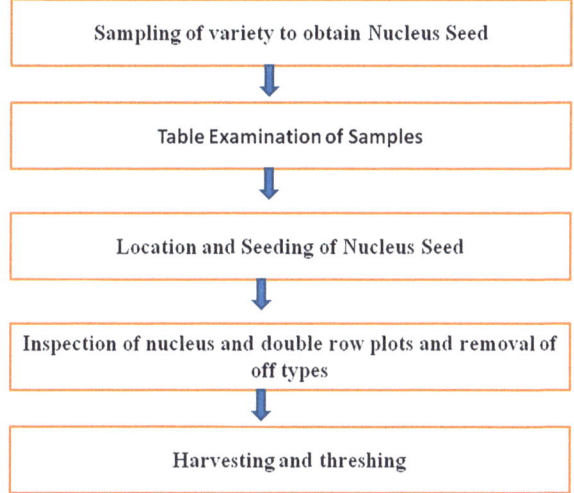

Fig. 1 A schematic representation of nucleus seed multiplication (Harrington, 1952)

The nucleus plot should be carefully checked at every stage of the growing season, from seedling to maturity. Differences in leaf colors, rate of development and growth patterns, heading times, plant height, and spike features should be examined with precision, and any plots that deviate clearly from the rest should be eliminated.

V. Threshing and harvesting

At least 180 of the original 200 lines in the nucleus seed production plot should be harvested individually and wrapped in various bundles. Each line's total bundles should be labeled and stored until the yield trials determine the line's future.

Each plot is cleaned individually and piled up on the seed table when the line has been identified or is about to be identified as a variety. Every plot is checked for a general homogeneity in seed appearance, and any deviation deemed unsatisfactory is eliminated. For the use in the following year, all of the remaining bundles are combined into one lot and preserved as "Breeder's Stock Seed."

Biofortification

Improving the nutrients in crops can be achieved through genetics, which involves utilizing the existing genetic diversity panel for classical breeding strategy and marker-assisted selection; besides, modern concepts of related gene discoveries are amenable (Grusak, 2002). Using plant breeding techniques, genetic biofortification creates staple food crops with higher amounts of micronutrients, lower levels of anti-nutrients, and higher levels of nutrients that aid in nutrient absorption (Bouis,

2003). By utilizing natural genetic variability to develop genotypes that are rich in mineral concentration, it provides a long-term solution to the issue of malnutrition (Pfeiffer & McClafferty, 2007). Plant breeders have examined current accessions in international germplasm repositories or collections to determine whether there is enough genetic diversity to breed for a certain characteristic. Then they have carefully developed nutrient cultivars of important staples that are high in Zn and Fe concentrations and have additives that increase Zn and Fe bioavailability, particularly in wheat.

(a) **Germplasm screening**: Significant genetic heterogeneity for grain Fe and Zn concentrations has been found in wheat and its wild relatives after germplasm screening. Landraces and wheat ancestors like einkorn wheat and wild emmer wheat have the greatest Zn and Fe values (Ortiz-Monasterio et al., 2007).

(b) **Breeding target and target population**: General estimations of breeding targets for Fe and Zn have been developed based on the themes of projected bioavailability percentage, daily consumption of wheat per person, method of food preparation, and estimated average requirements.

(c) **Breeding strategies**: Modern wheat breeding initiatives have focused mostly on raising yields in order to boost productivity. This has been accomplished in large part through selecting for critical qualities like disease resistance, short plant height, higher biomass, and harvest index (Trethowan et al., 2007).

(d) **Precision phenotyping**: pH, temperature, radiation, precipitation, organic matter, and soil texture are just a few of the environmental and soil factors that may have an impact on the concentration and solubility of micronutrients for plant roots (Joshi et al., 2010). Precision phenotyping is an essential technique to breed wheat germplasm with stable and high Zn concentrations.

(e) **High-throughput screening methodology**: Nutrient-rich genotypes must be quickly, precisely, and affordably identified in order to choose genotypes with higher micronutrient concentrations. The earlier method, Inductively coupled plasma-optical emission spectrometry (ICP-OES) (Zarcinas et al., 1987), is expensive, calls for highly skilled analysts, needs contamination-free reagents, and requires lengthy sample preparation. However, subsequently Atomic Absorption Spectrometry (AAS) (Mohammed, 2021), X-ray Fluorescence (XRF)-based non-destructive methods are widely used for elemental analysis (Borgwardt & Wells, 2017).

(f) **G X E interaction**: Soil composition with a lack of precisely essential microelements, including zinc, is a limiting factor in its improvement when the source itself is insufficient to meet the need, even with modern breeding procedures for efficient absorption or mobilization of the grain. Thus, Zn and Fe have shown significant genotype and location (environmental) relationships in both wild and cultivated wheat cultivars (Velu et al., 2012).

(g) **Associations between micronutrients and protein**: It would be easier to choose mineral-dense offspring through breeding with desired phenological and consumer-preferred features if we understood the relationships between micronutrients and grain production, plant height, grain size, and end-use quality factors. Research on wheat has shown that grain Zn and Fe are positively associated (Velu et al., 2011a, b, 2012). This suggests that the alleles responsible for Zn and Fe deposition in the grain tend to co-segregate or are pleiotropic, and as a result, Zn and Fe can be improved simultaneously.

(h) **Gene discovery**: Without having to measure mineral levels in field settings, we can choose advantageous genotypes by locating genetic markers connected to loci and determining variance for micronutrients. Quantitative trait loci (QTL) connected to Zn in wheat have only been discovered in a small number of studies. A rich allelic variety is present in wild emmer wheat, including for grain Zn and Fe contents (Xie & Nevo, 2008).

Biofortification in wheat is done mainly for:

Iron (Fe)

Due in part to their low bioavailability from plant food sources, which account for the majority of caloric intake in communities with limited resources, iron and zinc are two of the most often reported micronutrient deficits.

Iron-Deficiency Anemia (IDA) affects women of reproductive age and children under the age of five, according to the National Nutrition Survey. IDA has an impact on adult job productivity, children's physical and cognitive development, and maternal and child mortality.

Zinc (Zn)

Zn insufficiency is the most common micronutrient deficiency for many crops worldwide (Sadeghzadeh, 2013). Zn deficiency in plants has a significant impact on their production of pollen, growth, and the synthesis of carbohydrates, proteins, nucleotides, chlorophyll, auxin, cytochrome, and more (Hacisalihoglu & Blair, 2020). Depending on cultivars and the degree of soil Zn deficit, Zn shortage can reduce photosynthesis by up to 50–70%, which has an impact on yield components (Niyigaba et al., 2019).

Low human Zn assimilation is presumed to be a deficit in soil Zn content in wheat-growing regions, which results in naturally lower grain Zn concentrations. A short-term solution to the issue, agronomic biofortification (such as the use of Zn fertilizers), is preferable to the breeding strategy. However, foliar Zn applications provide notable increases in grain Zn content in wheat, whereas soil Zn applications are less successful at increasing grain Zn. Wheat grain Zn concentration might be further improved by maximizing the time and solute concentration of foliar Zn treatment, not only in whole grains but also in the endosperm.

The large grain accumulates more Zn, most likely when foliar Zn fertilizers are supplied to plants at a late growth stage, which is a significant element in evaluating the effectiveness of foliar Zn application in boosting grain Zn concentration.

Protein

Grain protein content (GPC) influences the nutritional value, processing preferences, end product quality (bread and pasta), and market value of both hexaploidy and tetraploidy wheat. An important goal of wheat breeding projects, especially for those aiming to improve nutritional quality, has been to increase GPC and change the composition of storage proteins in wheat grain because the economic worth of wheat grains depends on their GPC.

Development of Zn Rich Varieties in India

Realizing the prominence of the nutritional quality of wheat, research efforts were streamlined to the release or testing of 393 nutrient-rich crop varieties in 63 nations globally (Virk et al., 2021). In India, a total of 28 biofortified varieties have been notified, of which 12 have been developed as Zn-rich cultivars (Yadava et al., 2018; Kamble et al., 2022). These 12 Zn-rich varieties include four durum varieties. The Zn-biofortified wheat varieties have shown growth of 14.6–24.6% over the base level of 35 ppm of conventional bread varieties. These varieties are described in the following headings, Table 1.

Zn-Rich Varieties Recommended for Cultivation in India

1. WB02: Pedigree: T. DICOCCON CI9309/AE.SQUARROSA (409)/3/MILAN/S87230// BAV92/4/2*MILAN/S87230//BAV92. The variety has been notified for cultivation under timely sown conditions of North Western Plains Zone of India in 2017. With the Zn content of 42.0 ppm, it also contains high Fe (40.0 ppm) and average yield of 51.6 q/ha. (Chatrath et al., 2018).
2. HPBW01: Pedigree: T. DICOCCON CI9309/AE.SQUARROSA(409)/3/MILAN/S87230 //BAV92/4/2*MILAN/S87230 //BAV92. The variety has been notified for cultivation under timely sown conditions of North Western Plains Zone of India in the year 2017. With the Zn content of 40.6 ppm, it also contains high Fe (40.0 ppm) and average yield of 51.7 q/ha. (Yadava et al., 2018)
3. Pusa Tejas (HI8759): Pedigree: HI 8663/HI 8498 This durum variety with high grain Zn content (42.8 ppm) along with better protein (12%) and iron (42.1 ppm) concentration and has been recommended for cultivation in irrigated timely

Table 1 High-zinc wheat varieties were released between 2017 and 2021, exhibiting a high concentration of grain zinc

SN	Variety	Year of release	Zn content (ppm)	Gain over baseline level (35 ppm) (%)
1	WB 02	2017	42	20.0
2	HPBW01	2017	40.6	16.0
3	PUSA TEJAS (HI 8759) (D)	2017	42.8	22.3
4	HI 8777 (D)	2018	43.6	24.6
5	MACS 4028 (D)	2018	40.3	15.1
6	PBW 757	2018	42.3	20.9
7	PBW 771	2020	41.4	18.3
8	HI 1633	2020	41.1	17.4
9	DBW 327 (Karan Shivani)	2021	40.6	16.0
10	HI 1636	2021	40.4	15.4
11	HI 8823 (D)	2021	40.1	14.6
12	HUW 838	2021	41.8	19.4

sown conditions of Central Zone in the year 2017. The variety has average yield of 57.0 q/ha (Ambati et al., 2019)
4. HI8777 Pedigree: B931IID4672//HI8627. This durum variety with high grain Zn content (43.6 ppm) along with iron (48.7 ppm) concentration and has been recommended for cultivation in rainfed timely sown conditions of Peninsular Zone in the year 2018. The variety has average yield of the variety is 18.5 q/ha.
5. MACS4028 Pedigree: MACS 2846/BHALEGAON3*2. This durum variety with high grain Zn content (40.3 ppm) along with iron (46.1 ppm) and protein content (14.7%) concentration and has been recommended for cultivation in rainfed timely sown conditions of Peninsular Zone in the year 2018. The average yield of the variety is 19.3 q/ha.
6. PBW757 Pedigree: PBW55O/YR15/6*AVOCET/3/2*PBW550/4/PBW568+YR36/3*PBW550. This variety with high grain Zn content (42.3 ppm) has been recommended for cultivation very late sown irrigated conditions of North Western Plains Zone in the year 2018. The variety has average yield of the variety is 36.7 q/ha.
7. PBW771 Pedigree: BW9246/2*DBW71. This variety with high grain Zn content (41.4 ppm) has been recommended for cultivation under late sown irrigated conditions of North Western Plains Zone in the year 2020. The average yield of the variety is 50.3 q/ha.
8. HI1633 Pedigree: GW 322 / PBW 498. This variety has high grain Zn content (41.1 ppm) along with iron content (41.6 ppm) and protein concentration of 12.4% and it has been recommended for cultivation under late sown irrigated conditions of Peninsular zone in the year 2020. The average yield of the variety is 41.7 q/ha.

9. DBW327 Pedigree: NELOKI//SOKOLL/EXCALIBUR. This variety has high grain Zn content (40.6 ppm) and it has been recommended for cultivation under early sown irrigated conditions of North Western Plains Zone in the year 2021. The variety has average yield of the variety is 79.4 q/ha.
10. HI1636 Pedigree: DL788-2/HW4032. This variety has high grain Zn content (40.4 ppm) and it has been recommended for cultivation under timely sown irrigated conditions of central zone in the year 2021. The average yield of the variety is 56.6 q/ha.
11. HI8823 Pedigree: HI 8709/HD 4676. This durum variety has high grain Zn content (40.1 ppm) and it has been recommended for cultivation under timely sown restricted irrigated conditions of central zone in the year 2021. The average yield of the variety is 38.5 q/ha.
12. HUW838 Pedigree: This variety has high grain Zn content (41.8 ppm) and it has been recommended for cultivation under restricted irrigated conditions of North Western Plains Zone in the year 2021. The average yield of the variety is 51.3 q/ha.

Seed Production Programme of Zn-Rich Varieties

The Indian seed production program largely sticks with the limited generations system for seed multiplication in a phased manner, and the system recognizes the three generations model, namely breeder, foundation, and certified seeds, provides ample protection for quality assurance in the seed multiplication, and maintains the genetic and physical purity as it moves from the breeder to the farmer (https://seednet.gov.in).

In India, breeder seed is produced on the basis of consolidated indents from Department of Agriculture Cooperation (DOAC), Government of India, New Delhi which compiles the breeder seed requirements of the different public and private sector seed agencies. These organizations forecast the seed demand for various crop varieties in advance. The consolidated indent is forwarded to the Indian Council of Agricultural Research, which, through its vast network of various institutes and State Agricultural Universities (SAUs), carries out breeder seed production and fulfils the seed requirements of the different indenting agencies (Fig. 2). The Crop Science Division of the ICAR coordinates the breeder seed production of field crops in the country with the cooperation of DOAC. The responsibility of foundations and certified seed production vests with State and National Seed Cooperation and SAUs (Chauhan et al., 2017).

Though the research on biofortification was scattered throughout the country, a systematic approach was adopted after Harvest Plus program implementation in India in 2007 (Sanjeeva et al., 2020). At present, the Indian Council of Agricultural Research (ICAR), New Delhi, is playing a major role in ensuring food and nutritional security for the nation by providing the research thrust for the production of biofortified food grains. After the release of the first series of Zn-rich varieties in the

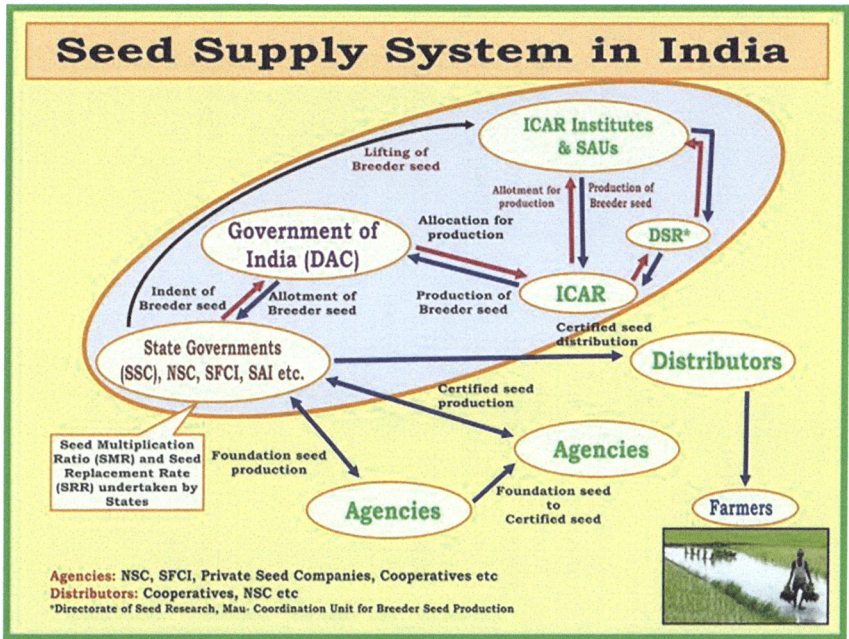

Fig. 2 An illustration of the seed supply system and associated networks in India

year, the breeder seed production of Zn-rich cultivars has been up scaled significantly. During the years 2017–18, a total of 364 quintals of breeder seed were indented, which has scaled up 2.7 times to the level of 996.4 q during 2021–22. Significant improvements in seed production of Zn-rich varieties have also been reported with 986 q during 2017–18 and reached 2204.7 q (2.2 times) during 2021–22. Further involvement of the private sector has enhanced the availability of these varieties to farmers at a faster rate (Neeraja et al., 2022). As per the reports, the recently released Zn-rich variety DBW327 has been licensed to 100 private seed growers for seed production and distribution. The targeted breeding for increased nutrient content has significantly impacted the development and deployment of biofortified wheat varieties in India (Kamble et al., 2022).

Strategies

National and International Support

Globally, nutrition enrichment through genetic improvement has been supported by Harvest Plus since 2003 in various food crops (Bouis & Saltzman, 2017) and has also contributed to the development of national biofortification breeding programs in Bangladesh, India, and Rwanda (Douthwaite, 2021). The Department of

Table 2 Breeder seed indent and production of Zn rich wheat varieties

SN	Variety	2017–18		2018–19		2019–20		2020–21		2021–22	
		Indent	Production	Indent	Production	Indent	Production	Indent	Production	Indent	Production
1	WB 02	37.00	500.30	192.40	193.00	358.80	501.00	80.20	90.00	112.00	129.70
2	HPBW01	75.00	100.00	152.60	153.00	169.20	170.00	191.00	240.00	177.60	220.00
3	HI 8759 (d)	252.00	386.00	323.20	649.50	356.40	550.00	846.20	420.00	638.80	800.00
4	HI 8777(d)					8.00	125.00	2.00	110.50	30.80	120.00
5	MACS 4028(d)					10.00	11.00	2.00	5.00		
6	PBW 757					37.00	40.00	33.00	33.00	12.80	30.00
7	PBW 771					0.00	1.00	71.80	77.00	17.40	25.00
8	HI 1633									7.00	130.00
9	DBW 327									0.00	750.00
10	HI 1636										
11	HI 8823(d)										
12	HUW 838										
	Total	364.00	986.00	668.20	995.5	939.4	1398.0	1226.2	975.50	996.4	2204.7

Biotechnology (DBT) and the Indian Council of Agricultural Research (ICAR) of the Government of India have also initiated support for biofortification projects, leading to consorted national and international research efforts towards the development of biofortified varieties (Sanjeeva et al., 2020). Intensive efforts by all national and international partners are facilitated by policies like mainstreaming biofortification in the national breeding programs (Ramadas et al., 2020). Further, the Zn main streaming efforts through CIMMYT target reducing the yield gap between biofortified and non-biofortified varieties and ensuring that 75% of the breeding line is nutri-rich (Govindan et al., 2022).

Government Support

The production and dissemination of good-quality seed and strengthening the seed chain to enhance the availability of biofortified varieties in different crops have been important steps to eliminating malnutrition (Yadava et al., 2018). Assurance of a remunerative price through the minimum support price and/or premium price for biofortified grains in the market will also encourage farmers to grow more biofortified crops. Recently unveiled National Nutrition Strategy 2017 by NITI Aayog, the Government of India envisages alleviation of malnutrition in the country through food-based solutions. To spread awareness of these biofortified cultivars among the general public, extra efforts are being made. For commercial production, high-quality seeds from biofortified cultivars are created and made available. The Extension Division of ICAR has established two unique projects to help scale up biofortified varieties through Krishi Vigyan Kendras (KVKs), namely Nutri-sensitive Agricultural Resources and Innovations (NARI) and Value Addition and Technology Incubation Centres in Agriculture (VATICA) (Yadava et al., 2018). Integration of biofortified varieties in government-sponsored programs such as the National Food Security Mission and Rashtriya Krishi Vikas Yojna, as well as nutrition intervention programs such as the Integrated Child Development Services scheme, 'Mid-Day Meal', and Nutrition Education and Training through Community Food and Nutrition Extension Units, will further enhance the health of the populace.

Development of High Yielding Biofortified Varieties

The improvement of people's general health is made possible via biofortification. These high-yielding biofortified varieties are more resistant to diseases, pests, droughts, etc. It provides a low-dose, food-based, sustainable substitute for iron supplementation. It could benefit farmers as well as the most vulnerable members of society who cannot afford food supplements. Since the technique can be easily repeated and scaled up after the initial research is completed, it is quite cost-effective. Additionally, through traditional methods employed in agriculture and the food industry, biofortified crops with higher bioavailable concentrations of vital micronutrients are delivered to consumers. This offers a practical means of reaching undernourished and low-income families with limited access to a variety of diets, supplements, and fortified foods. The development of biofortified crops eliminates the need to purchase fortificants and add them to the food supply during processing, making biofortification, from an economic perspective, a one-time investment that provides a cost-effective, long-term, and sustainable method of addressing hidden hunger.

Dynamic Seed System

The creation, distribution, and sale of certified seeds of "improved" varieties at authorized channels is referred to as the formal seed system. It typically only includes a small number of crops with higher economic values. National laws and rules regulating variety release, seed certification, and phytosanitary restrictions generally govern this system. Three crucial tasks that effective seed systems must accomplish include variety control, seed production, and seed distribution.

Faster Deployment of Varieties

Various organizational structures for seed production have evolved over time, based on the crop or variety, usage, market value, and quality criteria applied. These include the production of seeds by individual farmers, centralized certified seed production, and, more recently, various decentralized seed production schemes. One benefit of producing and storing seed on-site is that it is far less expensive than certified seed and is available to farmers at the time of planting. Although farmers have knowledge and expertise in seed selection, treatment, and storage, their seed production practices frequently contain flaws.

The majority of public seed producers are committed to producing certified seed of important food security crops, which are less appealing to the private sector than crops with a high multiplication factor (i.e., high net yield of seed per seeding rate). Private and parastatal firms produce seed on big farms and/or hire small- and medium-sized farmers as out-growers to scale up production with the goal of accessing national, regional, and/or international seed markets.

ICAR-IIWBR Model of Seed Production and Diffusion

The ICAR-IIWBR seed model comprises inter-institutional collaboration with the ICAR-National Dairy Research Institute, Karnal, the ICAR-Indian Institute of Farming System Research, Modipuram, and the ICAR-Central Potato Research Institute Regional Station, Modipuram. In total, 458 acres of land were under the seed production program under three ICAR institutes during 2022–23, which resulted in 9040.0 q of quality seed production of varied wheat varieties (Sharma et al., 2023). These institutes are engaged in quality seed production of various wheat varieties supplied by ICAR-IIWBR, Karnal, which is further buyback, processed, and packed at ICAR-IIWBR. All the precautions and quality regulation measures are being taken during seed production, harvesting, drying, processing, and storage to avoid any mechanical mixtures. Utmost care is taken during storage to avoid any losses due to storage insects and pests.

Licensing of varieties is one option that effectively utilizes the potential of private seed companies to make quality seeds available to farmers at a faster pace. More than 800 private seed companies have signed MOA with ICAR-IIWBR, Karnal, and enhanced the reach of wheat varieties to far-flung places at a faster rate.

The ICAR-IIWBR Seed Portal, designed and developed during the COVID-19 period, allows farmers to register their seed requirements online, and registered farmers are called for seed collection in groups or they are supplied seeds through postal or courier services. The portal has become quite popular among farmers, and in 2021–22, more than 20,000 farmers will have collected the seeds of recently released varieties.

Although the country has made significant progress in the development and deployment of Zn-rich varieties at a faster pace, there is still scope to differentiate the bio-fortified grains at the consumer level and demonstrate the impact of biofortified varieties on human health, especially among children and women.

References

Ambati, D., Prasad, S. S., Singh, J. B., Verma, D. K., Mishra, A. N., Prakasha, T. L., Phuke, R. M., Sharma, K. C., Singh, A. K., Singh, G. P., & Prabhu, K. V. (2019). High yielding durum wheat variety HI 8759 Pusa Tejas. *Indian Farming, 69*(04), 20–22.

Borgwardt, T. C., & Wells, D. P. (2017). What does non-destructive analysis mean? *Cogent Chemistry 3*(1), p. 1405767.

Bouis, H. E. (2003). Micronutrient fortification of plants through plant breeding: Can it improve nutrition in man at low cost? *Proceedings of the Nutrition Society, 62*, 403–411.

Bouis, H. E., & Saltzman, A. (2017). Improving nutrition through biofortification: A review of evidence from HarvestPlus, 2003 through 2016. *Global Food Security, 12*, 49–58.

Chatrath, R., Tiwari, V., Gupta, V., Kumar, S., Singh, S. K., Mishra, C. N., Venkatesh, K., Saharan, M. S., Singh, G., Tyagi, B. S., & Tiwari, R. (2018). WB 2: A high yielding bread wheat variety for irrigated timely sown conditions of North Western Plains Zone of India. *Wheat and Barley Research, 10*(1), 40–44.

Chauhan, J. S., Prasad, R. S., Pal, S., & Choudhury, P. R. (2017). Seed systems and supply chain of rice in India. *Journal of Rice Research, 10*(1), 9–16.

Douthwaite, B. (2021). Study on HarvestPlus' contribution to the development of national biofortifcation breeding programs. https://doi.org/10.2499/p15738coll2.134880

Gadwal, V. R. (2003). The Indian seed industry: Its history, current status and future. *Current Science, 84*(3), 399–406.

Govindan, V., Singh, R. P., Juliana, P., Mondal, S., & Bentley, A. R. (2022). Mainstreaming grain zinc and iron concentrations in CIMMYT wheat germplasm. *Journal of Cereal Science, 105*, 103473.

Grusak, M. A. (2002). Enhancing mineral content in plant food products. *Journal of the American College of Nutrition, 21*(sup3), 178S–183S.

Hacisalihoglu, G., & Blair, M. (2020). Current advances in zinc in soils and plants: Implications for zinc efficiency and biofortification studies. In *Achieving sustainable crop nutrition* (Vol. 76, pp. 337–353). Burleigh Dodds Science Publishing.

Harrington, J. F. (1952). Effect of variety and area of production on subsequent germination of lettuce seed at high temperatures. *Proceedings of the American Society for Horticultural Science, 59*, 445–450.

Joshi, A. K., Crossa, I., Arun, B., Chand, R., Trethowan, R., Vargas, M., & Ortiz-Monasterio, I. (2010). Genotype x environment interaction for zinc and iron concentration of wheat grain in eastern Gangetic plains of India. *Field Crops Research, 116*(268), 277.

Kamble, U., Mishra, C. N., Govindan, V., Sharma, A. K., Pawar, S., Kumar, S., Krishnappa, G., Gupta, O. P., Singh, G. P., & Singh, G. (2022). Ensuring nutritional security in India through wheat biofortification: A review. *Genes, 13*(12), 2298.

Komala, N. T., Gurumurthy, R., & Surendra, P. (2017). OMICS technologies towards seed quality improvement. *Indian Journal of Pure & Applied Biosciences, 5*, 1075–1085.

Mohammed, A. M. (2021). Elemental analysis using atomic absorption spectroscopy. *European Journal of Engineering and Technology Research, 6*(7), 48–51.

Neeraja, C. N., Hossain, F., Hariprasanna, K., Ram, S., Satyavathi, C. T., Longvah, T., Raghu, P., Voleti, S. R., & Sundaram, R. M. (2022). Towards nutrition security of India with biofortified cereal varieties. *Current Science, 123*(3), 271.

Niyigaba, E., Twizerimana, A., Mugenzi, I., Ngnadong, W. A., Ye, Y. P., Wu, B. M., & Hai, J. B. (2019). Winter wheat grain quality, zinc and iron concentration affected by a combined foliar spray of zinc and iron fertilizers. *Agronomy, 9*(5), 250. https://doi.org/10.3390/agronomy9050250

Ortiz-Monasterio, I., Palacios-Rojas, N., Meng, E., Pixley, K., Trethowan, R., & Pena, R. J. (2007). Enhancing the mineral and vitamin content of wheat and maize through plant breeding. *Journal of Cereal Science, 46*, 293–307.

Pfeiffer, W. H., & McClafferty, B. (2007). HarvestPlus: Breeding crops for better nutrition. *Crop Science, 47*, 88–105.

Ramadas, S., Vellaichamy, S., Ramasundaram, P., Kumar, A., & Singh, S. (2020). Biofortification for enhancing nutritional outcomes and policy imperatives. In *Wheat and barley grain biofortification* (pp. 309–327). Woodhead Publishing.

Sadeghzadeh, B. (2013). A review of zinc nutrition and plant breeding. *Journal of Soil Science and Plant Nutrition, 13*(4), 905–927.

Sanjeeva, R. D., Neeraja, C. N., Madhu, B. P., Nirmala, B., Suman, K., Rao, L. V. S., Surekha, K., Raghu, P., Longvah, T., Surendra, P., Kumar, R., Babu, V. R., & Voleti, S. R. (2020). Zinc biofortified rice varieties: Challenges, possibilities, and progress in India. *Frontiers in Nutrition, 7*, 26. https://doi.org/10.3389/fnut.2020.00026

Sharma, A. K., Kamble, U., Mishra, C. N., Kumar, S., Singh, G. P., & Singh, G. (2023). *IIWBR-seed model: Innovative outreach approach for accelerated gains in wheat* (p. 35). ICAR-Indian Institute of Wheat and Barley Research.

Smith, J. S. C., & Register, J. C., III. (1998). Genetic purity and testing technologies for seed quality: A company perspective. *Seed Science Research, 8*(2), 285–294.

Trethowan, R. M., Reynolds, M. P., Ortiz-Monasterio, I., & Ortiz, R. (2007). The genetic basis of Green Revolution in wheat production. *Plant Breeding Reviews, 28*, 39–58.

Velu, G., Ortiz-Monasterio, I., Singh, R. P., & Payne, T. (2011a). Variation for grain micronutrients concentration in wheat core-collection accessions of diverse origin. *Asian Journal of Crop Science, 3*, 43–48.

Velu, G., Singh, R. P., Huerta-Espino, J., & Peña, R. J. (2011b). Breeding for enhanced zinc and iron concentration in CIMMYT spring wheat germplasm. *Czech Journal of Genetics and Plant Breeding, 47*, S174–S177.

Velu, G., Singh, R. P., Huerta-Espino, J., Peña-Bautista, R. J., Arun, B., Mahendru-Singh, A., YaqubMujahid, M., Sohu, V. S., Mavi, G. S., Crossa, J., Alvarado, G., Joshi, A. K., & Pfeiffer,

W. H. (2012). Performance of biofortified spring wheat genotypes in target environments for grain zinc and iron concentrations. *Field Crops Research, 137*, 261–267.

Virk, P. S., Andersson, M. S., Arcos, J., Govindaraj, M., & Pfeiffer, W. H. (2021). Transition from targeted breeding to mainstreaming of biofortification traits in crop improvement programs. *Frontiers in Plant Science, 14*, 703990.

Witcombe, J. R., Packwood, A. J., Raj, A. G. B., & Virk, D. S. (1998). The extent and rate of adoption of modern cultivars in India, In J. R. Witcombe, D. S. Virk, & J. Farrington, (Eds.), *Seeds of choice: Making the most of new varieties for small farmers* (pp. 53–68). Intermediate Technology Publications.

Xie, W., & Nevo, E. (2008). Wild emmer: Genetic resources, gene mapping and potential for wheat improvement. *Euphytica, 164*, 603–614.

Yadav, R. N., Kumar, P. R., Hussain, Z., Yadav, S., Lal, S. K., Kumar, A., Singh, P. K., Bera, A., & Yadav, S. K. (2022). Maintenance breeding. In *Fundamentals of field crop breeding* (pp. 703–744). Springer.

Yadava, D. K., Hossain, F., & Mohapatra, T. (2018). Nutritional security through crop biofortification in India: Status & future prospects. *The Indian Journal of Medical Research, 148*(5), 621–631.

Zarcinas, B. A., Cartwright, B., & Spouncer, L. R. (1987). Nitric acid digestion and multielement analysis of plant material by inductively coupled plasma spectrometry. *Communications in Soil Science and Plant Analysis, 18*, 131–146.

Open Access This chapter is licensed under the terms of the Creative Commons Attribution 4.0 International License (http://creativecommons.org/licenses/by/4.0/), which permits use, sharing, adaptation, distribution and reproduction in any medium or format, as long as you give appropriate credit to the original author(s) and the source, provide a link to the Creative Commons license and indicate if changes were made.

The images or other third party material in this chapter are included in the chapter's Creative Commons license, unless indicated otherwise in a credit line to the material. If material is not included in the chapter's Creative Commons license and your intended use is not permitted by statutory regulation or exceeds the permitted use, you will need to obtain permission directly from the copyright holder.

Zinc Wheat Variety Release, Seed Production and Scaling Up Strategies in Pakistan

Muhammad Imtiaz

Introduction

More than two billion people, primarily in low- and middle-income countries, do not get enough essential vitamins and minerals (micronutrients) in their daily diets. This "hidden hunger" increases their vulnerability to serious health problems including weakened immunity, poor brain development, stunting, blindness, and even death (Rawat et al., 2013). Pakistan has a higher majority of stunted children than any other country in South Asia as reported in Asia and the Pacific Regional Overview of Food Security and Malnutrition (FAO, 2018). Four out of ten children under 5 years of age are stunted while 17.7% suffer from wasting, as per the National Nutrition Survey of Pakistan (NNS, 2018). The statistics for adolescents are also not very encouraging as about 41.7% of women of reproductive age (WRA) are anemic, with a slightly higher proportion in rural (44.3%) than in urban settings (40.2%) (NNS, 2018).

Pakistan grows ~9 million ha of wheat annually and is one of the highest wheat-consuming countries in the world as each Pakistani consumes, on average 240 g of wheat daily, or 87 kg annually, and thus provides 72% of Pakistan's daily caloric intake. This makes wheat the ideal food vehicle for an intervention to increase zinc intake in the population. Therefore, among potential interventions of fortification, and supplementation, HarvestPlus, CGIAR, and its partners introduced biofortification, which is the process of increasing micronutrients in the seed/grain of commercially available varieties. In this chapter, the focus is on Zinc (biofortified) wheat variety release, seed production, and scaling up strategies to increase the production and availability of biofortified grain for commercial production of high zinc flour, leading to improvement in the nutritional status of the population in the country.

M. Imtiaz (✉)
International Center for Agricultural Research in the Dry Areas (ICARDA), Islamabad, Pakistan
Previous affiliation: HarvestPlus, NARC, Islamabad, Pakistan
e-mail: m.imtiaz@cgiar.org

Biofortified Wheat Variety Release System

Current Varietal Release System

The varietal release system in Pakistan is very complex and sometimes time-consuming as both federal (central) and provincial (state) governments are involved in the variety release process. The Seed Act, 1976, and amended Act, 2015 govern varietal testing, registration, release, and approval systems in the country. Depending on breeding objectives, the wheat breeding programs in general employ three methodologies namely introduction, hybridization, and to a limited extent mutation to develop new wheat varieties with superior economic characteristics and disease resistance. Most of the wheat advanced lines/germplasm are introduced mainly from CIMMYT and ICARDA. The germplasm is tested for adaptability, regional suitability, and other important traits under diverse agroecological conditions. The elite lines are selected for further testing in yield trials at the national level as part of assessing value for cultivation and use (VCU).

Wheat breeding Programs use hybridization and modified pedigree/bulk selection to identify elite lines. After crossing two desirable parents, breeders study and observe their segregating populations (progenies) till F6–7 generations when most of the homozygosity is achieved. Afterward, the progenies are tested in replicated experiments with local/national variety as check-in Preliminary Yield Trials followed by Regular or Advanced Yield Trials, Micro yield trials (or provincial wheat yield trials). In mutation breeding M6 and M7 generations are evaluated. The best lines are harvested separately and tested in yield trials.

Briefly discussed below are the stages involved in the development, testing, registration, and release of a wheat variety with additional testing for micronutrients such as zinc in case of releasing a biofortified or varieties high in zinc contents (Fig. 1).

Preliminary Yield Trials (PYT-Year 1) Depending on the size of a breeding program, several lines (F6–F7 stage) selected from local hybridization program, exotic screening, and yield nurseries are tested in replicated trials with high-yielding commercial varieties as checks.

Advanced Yield Trials (AYT-Year-2) Based on performance in the PYTs, the selected lines/genotypes promoted to AYTs and are evaluated to compare performance with commercial checks.

Provincial/Regional Yield Trials (P/R YTs-Year 3) Before 2012, these trials were conducted in each province (state) with the name of micro-plot trials comprising selected genotypes/lines from AYTs. In partnership with national programs, we changed the nomenclature/names to Punjab Uniform Wheat Yield Trial (PUWYT) for Punjab province, Sindh Uniform Wheat Yield Trial (SUWYT) for Sindh province Khyber Pakhtunkhwa Wheat Yield Trial (KPUWYT) for Khyber

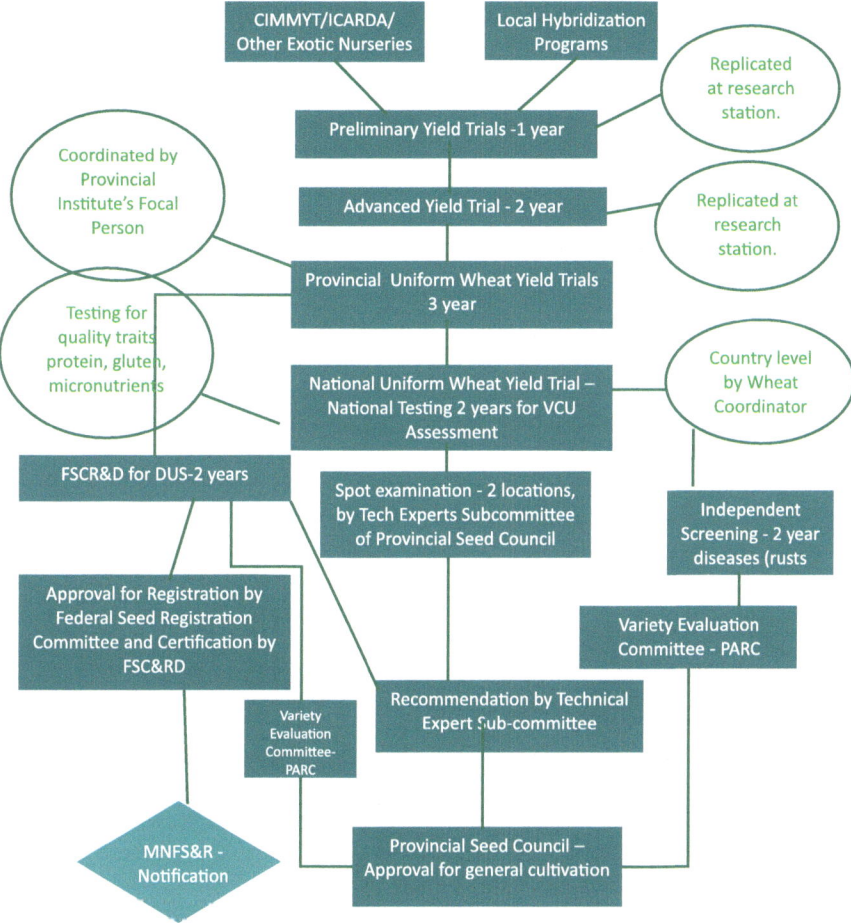

Fig. 1 Zinc (biofortified) variety development, testing, registration, and release system in Pakistan. Federal Seed Certification Department (FSCRD), Ministry of National Food Security and Research (MNFS&R), Pakistan Agricultural Research Council (PARC)

Pakhtunkhwa province, and Balochistan Uniform Wheat Yield Trial (BUWYT) for Balochistan province.

PUWYT Advanced lines from all wheat breeding programs in the province/or other interested public or private sector organizations/seed companies interested in releasing a variety in the province are submitted to the provincial focal person/institute i.e., the Director of Wheat Research Institute, Faisalabad, Punjab. The focal person coordinates the planting/testing of these advanced lines at 20–25 locations across the province. Provincial-level testing continues for at least 1 year and the selected lines/genotypes enter national-level testing for 2 years. It takes 5 years at provincial level testing before lines/genotypes enter the national level and DUS testing for 2 years.

Major provincial institutions contributing lines/genotypes to provincial and national level testing are Wheat Research Institute, Faisalabad, Reginal Agricultural Research Institute (RARI), Bahawalpur, Barani Agricultural Research Institute (BARI), Chakwal, Nuclear Institute for Agriculture and Biology (NIAB), Faisalabad, and Agricultural Universities/other interested parties from the private sector or central government institutions.

SUWYT Advanced lines from all wheat breeding programs in the province/or other interested public or private sector organizations/seed companies interested in releasing a variety in the province are submitted to the provincial focal person/institute i.e., While Wheat Research Institute, Sakrand, Sindh. The focal person coordinates the planting/testing of these advanced lines at 8–10 locations across the province. Provincial-level testing continues for at least 1 year and the selected lines/genotypes enter national-level testing for 2 years. It takes 5 years in provincial testing before elevating variety to national testing. The best performers in SUWYT are sent to National Uniform Wheat Yield Trials (NUWYT) and Distinctness, Uniformity, and Stability (DUS) testing for 2 years. After clearing NUWYT and DUS testing, the proposal is submitted to the Technical Sub Committee and finally submitted to the Sindh Seed Council for approval as a variety. The council grants permission and approves a variety for planting in Sindh Province on a commercial basis.

Major provincial institutions contributing lines/genotypes to provincial and national level testing are Wheat Research Institute, Sakrand, Nuclear Institute of Agriculture (NIA), Tandojam, Quaid e Awam Agriculture Research Institute Larkana (QAARI), Larkana, and Agricultural Universities/other interested parties from the private sector or central government institutions.

KPUWYT Advanced lines from all wheat breeding programs in the province/or other interested public or private sector organizations/seed companies interested in releasing a variety in the province are submitted to the provincial focal person/institute i.e., Director (Research Planning), Agriculture Research System, KP. The focal person coordinates the planting/testing of these advanced lines at 15–20 locations across the province. Provincial-level testing continues for at least 1 year and the selected lines/genotypes enter into national-level testing for 2 years. It takes 5 years at provincial level testing before lines/genotypes enter into national level and DUS testing for 2 years.

Major provincial institutions contributing lines/genotypes to provincial and national level testing are Cereals Crops Research Institute (CCRI) Pirsabak, Nowshera, Agriculture Research Institute (ARI), Tarnab, Peshawar, Agriculture Research Institute ARI), Dera Ismail Khan, Barani Agriculture Research Station (BARS), Kohat and Agricultural Universities/other interested parties from the private sector or central government institutions.

BUWYT Advanced lines from all wheat breeding programs in the province/or other interested public or private sector organizations/seed companies interested in releasing a variety in the province are submitted to the provincial focal person/institute i.e., Director Cereal Crops, Agriculture Research Institute, Sariab, Quetta, Balochistan. The focal person coordinates the planting/testing of these advanced lines at 4 ecological zones across the province. It takes 4 years in provincial testing commencing from the preliminary, advanced, micro plot and Baluchistan wheat yield trials before elevating a candidate variety to national testing. Provincial-level testing continues for 1/2 years and the selected lines/genotypes enter national-level testing for 2 years. It takes 5 years in provincial testing before elevating variety to national testing. The best performers in BUWYT are sent to NUWYT and DUS testing for 2 years.

The best-performing candidate varieties/lines in the provincial testing against checks are promoted to NUWYT.

NUWYT The national level focal person i.e. National Wheat Coordinator, Pakistan Agricultural Research Council (PARC) conducts a nationwide evaluation of candidate varieties/lines in diverse agroecological zones to generate VCU data for 2 years (Piepho et al., 2016). The Data from these trials are vital and mandatory for the National Breeding Programmes and variety release system. Trials are conducted at public sector research farms with all participating candidate varieties/lines coded by the Federal Seed Certification and Registration Department (FSC&RD) as an independent regulator. Before 2012, there were several types of NUWYTs such as irrigated, rainfed, late planting, and normal planting. However, with the author led national level brainstorming among national scientists, NUWYTs were redefined and only one NWUYT with an increasing number of entries and planting across both irrigated and rainfed conditions was finalized. This provided opportunities for breeders not only to select varieties for irrigated or rainfed conditions but also to identify those candidate varieties performing under both conditions. Additionally, the same NUWYT was also eligible for planting under late/early conditions depending on the requirement of a breeding program.

NUWYT-1st Year as stated above, candidate varieties showing promising results in provincial testing are promoted to NUWYT testing for agronomic value or VCU. These trials as described are multi-locational and coded ones planted at more than 50 locations throughout Pakistan.

NUWYT-2nd Year Candidate varieties performed better than local/regional and national checks are promoted to 2nd year of NUWYT testing to complete mandatory data requirements for the release of a variety.

Parallel to NUWYT, with slight variation in the provinces, there are other mandatory testing requirements such as disease screening, DUS testing, and spot examination of candidate varieties as briefly discussed below:

National Wheat Disease Screening For disease screening, all entries of NUWYT are provided to the Director, Crops Diseases Research Institute (CDRI), National Agricultural Research Centre (NARC) along with parentage and pedigree for National Wheat Disease Screening Nurseries, planted at 14–18 disease hotspot locations of the country. The independent report on disease response/incidence for each entry with emphasis on three rusts is produced and shared with all stakeholders including breeders and Wheat coordinator. Diseases screening is conducting for 2 years both under field and artificial epiphytotic conditions to avoid any escape.

DUS Trials During NUWYT testing, breeders also submit seed samples to FSC&RD to conduct DUS trials at one location for 2 years. The DUS data generated is provided to respective breeders before the provincial seed council meeting.

Testing for Zinc content (micronutrients) and Quality Traits In addition to quality data for the 1000-grain weight (TGW), test weight, starch, grain protein, and gluten contents, it is mandatory to provide micronutrients like zinc, iron data for labeling a variety as biofortified or high in zinc or other micronutrients.

Spot Examination The technical experts sub-committee (TESC) comprising representatives from FSC&RD, Pakistan Agriculture Research Council, Public Seed Corporation, and DGs Agri. Ext, Research, and representatives from Agricultural Universities, Federal Institutes, Directorate of Pathology, Virology, Entomology, and Agronomy, etc. The committee is mandated to check/evaluate field performance, and disease response, and review all the results of field trials conducted on the candidate varieties, especially concerning yield performance and responses to diseases, insects, and pests.

After spot examination of a candidate variety, the comprehensive proposal for variety is submitted to the expert subcommittee of provincial seed councils, comprising heads of each institution, representatives of public seed corporations/organizations, pathologists/entomologists from research institutes, and FSC&RD. The variety proponent breeder presents yield, disease, agronomic performance along with quality traits data, and NUWYT performance data to the expert subcommittee. The breeder planning to claim biofortified status for his/her variety needs to provide micronutrients such as zinc and iron contents data. After committee consensus and approval, candidate varieties are further submitted to respective Provincial Seed Councils for approval. and are commercially available to farmers after approval is granted. However, recently introduced changes in the varietal release system, the spot examination step has been eliminated and PARC led Varietal Evaluation Committee (VEU) recommendations were made mandatory.

Variety Approval and Release: Provincial Seed Councils

The concerned breeder presents all performance data and associated reports of a candidate variety to the provincial seed council. Provincial Seed Councils (Seed Act, 1976) chaired by Provincial Agriculture Ministers and Managing Directors of Seed Corporations that exist such as in Punjab and Sindh provinces, Secretary of Agriculture, Chairmen, PARC, FSC&RD, Extension Departments, and two to three progressive growers grant final approval for the release of a variety for commercial cultivation in the target/respective province. After formal approval of variety by the provincial Seed Council and registration by the Federal Seed Registration Committee, a variety is notified in the official gazette under clause 10 of Seed (Act) 1976, along with its area of adaptability and minimum limits for germination and purity for production of different classes of certified seed. Based on zinc level, till December 2023, five high zinc (biofortified) varieties were released in the country. Three varieties namely Zincol-2016, Akbar-2019, and Nawab-2021 were approved by Punjab Seed Council for Punjab Province while Tarnab-Rehbar and Tarnab-Gandum-I were approved by KP Seed Council for KP Province.

Proposed Modified Varietal Release System Despite mandatory testing in the NUWYT throughout Pakistan, the varietal release system is still based on administrative boundaries of the provinces (states) where each provincial seed council releases variety for a respective province i.e. Punjab, Sindh, KP, and Balochistan. Thus, significant investment in NUWYT data generation becomes senseless as each province only focuses on data points for testing locations in the province rather than all over the country.

In most countries worldwide, the varietal release system is based on agro-ecologies, for example, in India, the whole country is divided into six zones namely the Northern Hills Zone (NHZ), Northwestern Plains Zone (NWPZ), Northeastern Plains Zone (NEPZ), Central Zone (CZ), Peninsular Zone (PZ), and Southern Hills Zone (SHZ). Therefore, the rethinking of the varietal release system is a must where the country needs to move toward a release system based on agroecologies (Piepho et al., 2016) rather than administrative boundaries which will not only enable farmers to get more benefits from new genetics (including nutrients enriched varieties) but also help the country to increase its production and improve the nutritional status of the population.

In addition, the country is carrying the burden of malnutrition, therefore, to develop nutrient-enriched varieties, it should be mandatory to present micronutrients like zinc, iron, etc. data at the time of variety approval and mainstreaming of biofortification must be encouraged. To implement the proposed revised/modified release system, the legal entities at the Federal level, the Variety Evaluation Committee (VEC) and the National Seed Council (Seed Act, 1976, 2015) must encourage and enforce agroecologies based variety release; thus, enabling the regulator, FSC&RD to provide certification coverage to wheat seed crop in a particular ecology.

Current Wheat Seed Production System

The Pakistan wheat system is mostly supply driven and follows the limited generations' system for seed multiplication which recognizes three generations namely Pre-basic, Basic, and certified seeds. The regulator FSC&RD enforces adequate safeguards to ensure quality in the seed multiplication chain to maintain the genetic purity of the variety as it passes from the breeder to the farmer.

Pre-basic is the progeny of the breeder nucleus seed (BNS) of a variety and is produced by the breeder who developed the variety or his/her nominee. Pre-breeder seed production is the mandate of the provincial and federal agriculture research institutes/universities that developed and released the variety. The FSC&RD issued a "white tag with violet diagonal line" for this category/class of seed to differentiate it from other seed categories. Basic Seed is the progeny of pre-basic seed. Before the Seed Act 2015, it was mandatory to produce basic seed on government farms only, however, now the basic seed is produced by private seed-producing companies, agricultural extension departments, or seed corporations such as Punjab and Sindh Seed Corporations with a clear traceability to pre-basic and BN seeds. A white tag is issued to this category of seed. Certified seed which is the progeny of basic seed and is produced by both the Public and Private sectors and is recognized by the blue tag in the marketplace. Under certain circumstances, the approved category of seed that is produced from certified seed and meets the standards of seed is recommended with a pink label on it.

The Pakistani seed industry used to be dominated by public sector seed corporations and provincial Agricultural Research and Extension Departments as an exclusive provider of wheat seed. However, with the introduction of an enabling policy that eased government regulations, the seed industry was declared a business. Resultantly the private seed sector was encouraged to complement the efforts of the public sector. For 9 million hectares of wheat, approximately ~1 million certified seed is required; however, the seed availability fluctuates between 20% and 25% with the gap between the requirement and availability being most of the time over 70% (Hussain et al., 2017). Out of the available seed, ~80% certified seed market share is attributed to the private sector while ~20% of seed is provided by the public sector.

Scaling Strategies for Nutrition-Sensitive Seed System

Seed plays a major role in enhancing productivity and climate resilience, thus improving food and nutrition security and farmers' livelihoods. Therefore, an effective and functional seed system must be able to offer quality seeds of farmers' preferred adapted varieties for every planting season. However, Pakistan's seed sector faces several challenges (IFPRI, 2022) such as a supply and demand gap, a lengthy varietal approval process, a strict regulatory framework, no enforcement of Plant

Breeders' rights yet, and the presence of a large informal seed sector especially in the wheat crop. One of the reasons is that government efforts to improve seed systems are mainly focusing on the development of the formal seed sector and did not make any efforts to provide an enabling policy environment to promote the informal and integrated seed system approaches.

Foley et al. (2021) reviewed scaling up the delivery of biofortified staple food crops globally while HarvestPlus (HP, 2022a) presented a delivery and commercialization model for scaling nutrient-enriched crops. The key strategies to scale high zinc (biofortified) wheat in Pakistan are aligned with these proposed delivery models and the Journey of Scaling Zinc Wheat in Pakistan (HP, 2022b) with some modifications to emphasize the pre-release seed multiplication. Components of scaling high zinc (biofortified) wheat in Pakistan include investment in pre-released seed multiplication of potential high zinc candidate varieties, early generation seed multiplication enabling sizeable production of certified seed for commercial farming, seed demonstration, and popularization/demand generation of zinc enriched varieties and facilitating key partnership across the wheat seed value chain (Fig. 2).

Scaling Up Seed of Biofortified Wheat Varieties

Pre-released Seed Multiplication

The precursor to rapid adoption and scaling of biofortified wheat varieties is the pre-release seed multiplication, which refers to the multiplication of the seed of a variety before its formal release for commercial cultivation. This assures that at the

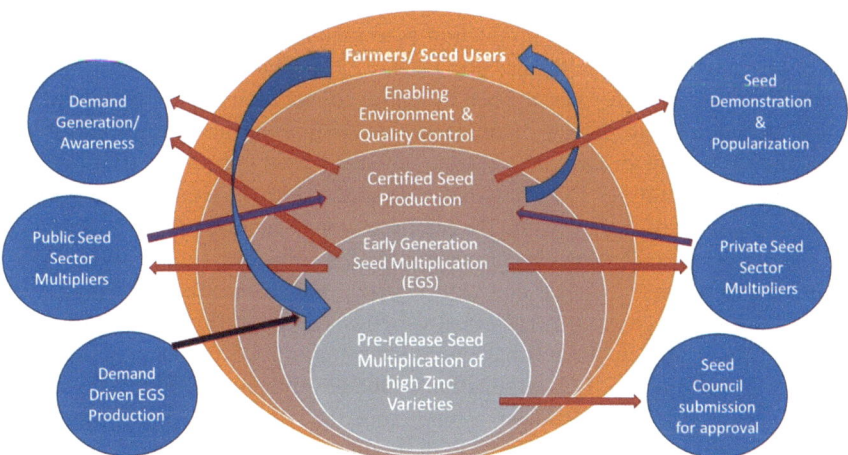

Fig. 2 Scaling model for nutrition sensitive (high zinc) wheat seed system

time of variety release, enough EGS (pre-basic) seeds are available for large-scale multiplication and rapid dissemination to farmers. Either due to statute restrictions or production costs, the amount of breeder seed available at the time of varietal release ranges from 100 to 500 kg which is too small to allow rapid multiplication of subsequent seed generations. This is leading to an increasing gap in varietal release and its adoption by farmers because as soon as farmers learn about the release of a new variety then there is a demand, but no seed is available in the market. (Joshi & Braun, 2022) explained the benefits of pre-release seed multiplication.

Harvestplus (HP) as part of its scaling strategy, invested in pre-release and EGS multiplication and continues to partner with public sector biofortified varietal developer organizations to make available EGS to public and private sector seed multipliers. For example, out of five biofortified wheat varieties namely Zincol-2016, Akbar-2019, Nawab-2021, Tarnab-Rehbar and Tarnab-Ghandum-1, released so far in Pakistan, the last three varieties have very small quantities of EGS at the time of release. Harvestplus supported the EGS production of these varieties, however, for sustainability, it must be mandatory to invest in the pre-release seed multiplication to reduce the gap when a variety is released and its adoption. At the same time the regulator i.e. FCSR&RD must stop the sale of seed of pre-release varieties in case there is a leakage of seed before its release to either farmers or private sector seed multipliers.

Post-release Seed Multiplication

Acknowledging the important role of pre-release seed multiplication in the scaling of biofortified varieties, the other key component in scaling is the sizable production of EGS. The purpose of EGS production is to maintain the genetic potential and identity of a variety and the regular provision of high-quality breeder seed for subsequent seed production. Quality seed of improved and high zinc varieties is difficult to access at scale due to bottlenecks in the EGS value chain. One of the constraints to having an early impact from newly released improved varieties is the inadequate availability of EGS required to produce certified seed to be grown by farmers. Thus, there is a need for the public sector to increase the production of early-generation seed for use by the private sector and other seed multipliers for further multiplication as basic and certified seed which is crucial for achieving impact from improved biofortified wheat varieties. Although HarvestPlus, CGIAR, and other donor-funded projects have been investing in EGS production, however, the institutionalization of EGS production by the government is a must for sustainability. The government of Pakistan recently took the initiative in this direction and provided significant funds to increase EGS production on a sustainable basis.

Transfer of Seed to Farmers

The sizeable production of certified seed for commercial farming ensures the availability of and access to quality seed of new biofortified wheat varieties which play a significant role in increasing agricultural production and productivity and ensuring food and nutritional security. Hussain et al. (2017) reviewed the wheat seed sector in Pakistan where the use of certified seed ranges from 20% to 25% while a large quantity of seed sown comes from either farmers' saved seed or seed obtained from fellow farmers, middlemen, or village shops. Slightly variable annually, the major share (14–18%) of certified seed comes from the private sector. Thus, the gap between the seed requirement and availability is continuously over 70%. Therefore, there is significant scope for increasing this percentage if we can produce demand-driven larger quantities of certified seed at affordable prices and streamline the dissemination of quality seeds among farmers which is discussed in the coming sections. Other key drivers to fast-track deployment and scaling of new biofortified wheat varieties in addition to pre-release and EGS production include wheat seed value chain partnership and the introduction of demand-driven seed systems.

Wheat Seed Value Chain Partnership

To scale up biofortified wheat seed, public-private partnerships across the seed value chain and the use of digital technology has significant scope to link formal and informal seed systems and network producers and markets thus ensuring the availability of the latest and improved biofortified wheat seeds in sufficient quantity catering the demand of the farmers. HarvestPlus and CGIAR established partnerships with public and private seed multipliers, and nongovernmental organizations, established seed demonstration plots as part of the varietal popularization/demand generation, and capacity-building of seed value chain actors which facilitated the rapid scaling and commercialization of biofortified wheat seed in Pakistan (Fig. 3).

Demand Driven Seed System

To scale farmers' preferred and nutrient-enriched varieties, structural reforms are required to establish a demand-driven wheat seed supply system replacing the current supply-driven system. At present there is no systematic way to identify the demand for new varieties and there is always a shortage of newly released improved seeds during the early years of a newly released variety. Private seed companies assess demand through informal feedback from researchers, agro-dealers, extension agents, and farmers' queries. Thus, the lack of information on demand means that seed companies do not produce enough seed of new improved varieties such as biofortified wheat and continue to promote the established and popular varieties, which are in most cases old and have a high risk of becoming susceptible to diseases such as rusts. Therefore, to meet the demand of the farmers and markets, the

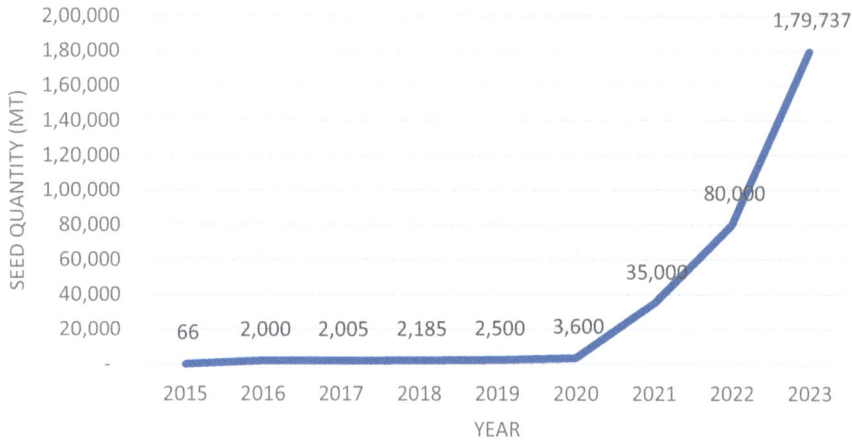

Fig. 3 Quantity of biofortified wheat seed produced (MT) in Pakistan

proposed seed system needs to make sure that varieties released within 5 years are promoted and old susceptible ones are eliminated which will require restructuring the current wheat seed system to an indent-based seed production system.

Out of two Seed Corporations in Punjab and Sindh, only Punjab Seed Corporations is functional while the extension departments in KP and Balochistan provinces are handling seed procurement and distribution/marketing. Under the current system, EGS flows from public sector varietal developers to seed multipliers both in the public and private seed sectors without any systematic demand assessment as discussed above.

Seed Indent-Based System

To scale nutrients enriched wheat seed, demand-based seed indents from various seeds producing entities must be introduced. Aligned with the proposed agroecology-based varietal release system in this chapter, every province shall provide the agro-ecology zone-wise and variety-wise quantity of certified seeds sold and area covered in the previous wheat sowing and EGS requirements for the coming season of each variety to either provincial Seed certification and registration department and or Agriculture Research Department in each province. Each province in consultation with FSCR&RD, shall formulate a seed plan (EGS and Certified seed) for the cropping seasons based on the performance of existing and new varieties. Both private and public seed multipliers/companies shall place the EGS seed indent a year in advance. The further refinement of this basic structure presented will require consultation and buy-in from all stakeholders working on wheat seed production and scaling in the country, including policymakers.

References

FAO. (2018). *Asia and the Pacific regional overview of food security and nutrition 2018 – Accelerating progress towards the SDGs, Bangkok.* License: CC BY-NC-SA 3.0 IGO.

Foley, J. K., Michaux, K. D., Mudyahoto, B., Kyazike, L., Cherian, B., Kalejaiye, O., Ifeoma, O., Ilona, P., Reinberg, C., Mavindidze, D., & Boy, E. (2021). Scaling up delivery of biofortified staple food crops globally: Paths to nourishing millions. *Food and Nutrition Bulletin, 42*(1), 116–132.

Harvestplus (HP). (2022a). *Scaling nutrient enriched crops with the HarvestPlus delivery and commercialization model.*

Harvestplus (HP). (2022b). *The journey of scaling in Pakistan – Zinc-enriched biofortified wheat.*

Hussain, A., Ali, A., & Imtiaz, M. (2017). Status of wheat seed sector in Pakistan: Opportunities and challenges to improve quality seed supply for increased production and food security. *Journal of Biology, Agriculture and Healthcare, 7*(14).

IFPRI. (2022). *PACE policy paper: Seed industry in Pakistan: Policy challenges and prospects.*

Joshi, A. K., & Braun, H. J. (2022). Seed systems to support rapid adoption of improved varieties in wheat. In M. P. Reynolds & H. J. Braun (Eds.), *Wheat improvement.* Springer. https://doi.org/10.1007/978-3-030-90673-3-14

National Nutritional Survey (NNS). (2018). Nutrition Wing, Ministry of National Health, Services, Regulations and Coordination, Pakistan.

Piepho, H.-P., Nazir, M. F., Qamar, M., et al. (2016). Stability analysis for a countrywide series of wheat trials in Pakistan. *Crop Science, 56*, 1–11.

Rawat, N., Neelam, K., Tiwari, V. K., & Dhaliwal, H. S. (2013). Biofortification of cereals to overcome hidden hunger. *Plant Breeding, 132*(5), 437–445.

Seed Act, 1976. (1976). *An Act to provide for controlling and regulating the quality of seeds of various varieties of crops.*

Seed (Amendment) Act. (2015).

Open Access This chapter is licensed under the terms of the Creative Commons Attribution 4.0 International License (http://creativecommons.org/licenses/by/4.0/), which permits use, sharing, adaptation, distribution and reproduction in any medium or format, as long as you give appropriate credit to the original author(s) and the source, provide a link to the Creative Commons license and indicate if changes were made.

The images or other third party material in this chapter are included in the chapter's Creative Commons license, unless indicated otherwise in a credit line to the material. If material is not included in the chapter's Creative Commons license and your intended use is not permitted by statutory regulation or exceeds the permitted use, you will need to obtain permission directly from the copyright holder.

Current Status of Zinc-Biofortified Rice Cultivation in Bangladesh

Khondoker Abdul Mottaleb, Alvaro Durand-Morat, Fazleen Abdul Fatah, and Md. Abdur Rouf Sarkar

Introduction

The rapid growth of gross domestic product (GDP) and agricultural productivity since the 1990s have tremendously contributed to the fight against hunger and poverty in Bangladesh. During 1990–2022, the annual average growth rate of agriculture, forestry, and fishing was 3.68%, while GDP grew 5.7% per year (World Bank, 2022). Consequently, Bangladesh's per capita GDP increased by more than 800% from US $295 in 1990 to US $2688 in 2022 (World Bank, 2022). Presently, the country has successfully upgraded itself to a lower-middle-income country. Also, the poverty rate in the country has decreased drastically. For example, 33.4% of the total population of Bangladesh was extremely poor in 2002, living on less than US $1.90 per day, which dropped to 3.4% in 2022 (Development Initiatives, 2020). In 1990, the daily per capita dietary energy intake rate was 2111 kcal, which increased to 2575 kcal/daily in 2022 (FAOSTAT, 2022b). In 2022, Bangladesh was ranked 84th out of 121 countries on the Global Hunger Index (Von Grebmer et al., 2022),

---

K. A. Mottaleb (✉)
Department of Agricultural and Applied Economics, Texas Tech University, Lubbock, USA
e-mail: k.mottaleb@ttu.edu

A. Durand-Morat
Department of Agricultural Economics and Agribusiness, University of Arkansas, Fayetteville, AR, USA

F. A. Fatah
Universiti Teknologi MARA (UiTM), Faculty of Plantation and Agrotechnology, Kuala Lumpur, Malaysia

M. A. R. Sarkar
School of Economics, Zhongnan University of Economics and Law, Wuhan, China

Agricultural Economics Division, Bangladesh Rice Research Institute, Gazipur, Bangladesh

© The Author(s) 2025
M. Govindaraj et al. (eds.), *Breeding Zinc Crops for Better Human Health*, https://doi.org/10.1007/978-3-031-84342-6_6

compared to 2006, when it was ranked 102nd out of 119 countries in terms of global hunger status (DWHH and IFPRI, 2006).

Despite the economic progress, in 2019, 28% of children under 5 years of age are stunted, and 9.8% of them are wasted (Development Initiatives, 2020). In addition, 36.7% of women of reproductive age are anemic, and 13% of the total population of the country is undernourished. Studies stressed that 41–44.6% of pre-school-aged children and 57.3% of non-pregnant and non-lactating women in Bangladesh are zinc deficient (Zn) (Rahman et al., 2016). According to a report, on average, 55% of the total population of Bangladesh is zinc deficient (Nutrition Connect, 2019). Zinc deficiency is directly associated with stunted linear growth and declined immune function (Rahman et al., 2016). Widespread and uncontrolled malnutrition, including anemia, can severely impact human capital formation, quality labor supply, and the country's long-run economic progress. It is estimated that the per capita income penalty only from childhood stunting is around 7% (Galasso & Wagstaff, 2019). Thus, while Bangladesh is highly successful in fighting against abject hunger, hidden hunger in the form of malnutrition is still a significant policy concern.

To fight hunger and hidden hunger (malnutrition) in the most cost-effective way, the government of Bangladesh has been trying to develop and scale out biofortified staple crops, such as rice and wheat. Crop biofortification is a process through which the concentration of essential and critically important vitamins and minerals in staple crops are enhanced through the breeding process (Saltzman et al., 2013). With a yearly per capita rice consumption of 257 kg, Bangladesh has the largest rice per capita consumption in the world, and in terms of total rice consumption, Bangladesh is the fourth largest rice-consuming country (42 million metric tons, MMT) in the world (FAOSTAT, 2022b). In Bangladesh, rice supplies daily 1711 kcal of dietary energy per person, which is 66% of the total daily dietary energy intake (FAOSTAT, 2022b). As the diet of Bangladesh is predominantly rice-based and as rice generally lacks some key vitamins and minerals, the risk of inadequate zinc intake is high among Bangladeshi citizens (Rahman et al., 2016). Thus, to fight zinc deficiency in the most effective way, the government aims to develop and disseminate zinc-biofortified rice in Bangladesh.

In 2013, the first zinc-biofortified rice, BRRI dhan62, was released in Bangladesh (BRRI, 2023). The variety can supply zinc at 19 milligrams per kilogram. Until 2023, the Bangladesh Rice Research Institute (BRRI) subsequently released a total of seven zinc-biofortified rice varieties, including the BRRI dhan62. In addition, Bangabandhu Sheikh Mujibur Rahman Agricultural University developed and released BU dhan2, a zinc-fortified rice, in 2016, and the Bangladesh Institute of Nuclear Agriculture (BINA) developed and released Binadhan-20, a zinc-enriched rice, in 2017 (Bashar, 2018).

In Bangladesh, there are three rice seasons. The wet season is the aman rice season (late June–January), the summer season is aus rice (mid-March–August), and the dry season is boro rice (November–June). Boro is the major season in Bangladesh, contributing around 50% of the total rice produced in the country (BBS, 2023a). Considering the importance of the boro season, five out of seven varieties developed by BRRI during this season, and only two biofortified rice varieties, BRRI dhan62 and BRRI dhan72, were released targeting the aman season (BRRI, 2023).

A few studies examined the efficacy of zinc-biofortified rice in addressing zinc deficiency. A controlled study conducted under the vulnerable group feeding program from January to December (Ara et al., 2019) indicated that the prevalence of anemia and zinc deficiency was lower among vulnerable women who consumed zinc-biofortified rice compared to their counterparts who consumed ordinary rice.

Still, another randomized controlled trial study found that the consumption of zinc-biofortified rice had no impact on zinc deficiency status in the experiment group of children aged 12–36 months; however, it favored the experiment group with a height-for-age z-score (Jongstra et al., 2022). Scientists and specialists argued that proper and regular consumption of zinc-biofortified rice could contribute to reducing anemia and increasing serum zinc. Experts and scientists stressed that proper and regular consumption of zinc-biofortified rice can meet 60% of the daily zinc demand of a person (Dhaka Tribune, 2023; Noyon, 2023).

While it is imperative to scale out zinc-biofortified rice in Bangladesh, despite the efforts of the public and private sectors, the adoption of zinc-biofortified rice is low in Bangladesh. For example, in 2018, only 1% of the total rice area of Bangladesh was planted with zinc-biofortified rice (Nutrition Connect, 2019). According to HarvestPlus (2017), around one million rice farmers in Bangladesh are cultivating biofortified crops. The zinc-biofortified rice varieties are a new technology for farmers in Bangladesh. The introduction of a new technology, such as a new seed, often involves uncertainty as the cost and benefits of adopting a new technology are unknown to farmers (Ding et al., 2023). This is why the adoption of any new agricultural technology is seldom rapid or instantaneous (e.g., Rogers, 1983; Feder et al., 1985; Byerlee & Polanco, 1986; Waller et al., 1998; OECD, 2001; Sunding & Zilberman, 2001; Foster & Rosenzweig, 2010; Suri, 2011). A question arises as to how to encourage farmers to cultivate zinc-biofortified rice in Bangladesh.

Every new technology or its derivatives, such as seeds in agriculture, is associated with risks. Thus, providing information about the potential benefits and risks of adopting a new technology can help shape farmers' perceptions of the new technology, which can facilitate a rapid adoption of it (Adesina & Zinnah, 1993; Adesina & Baidu-Forson, 1995; Negatu & Parikh, 1999). In the case of zinc-biofortified rice in Bangladesh, which is relatively new, it is important to ensure the provision of information to the farmers about the agronomic characteristics as well as the health benefits of consuming zinc-biofortified rice to encourage farmers to cultivate and consume the zinc-biofortified rice.

Empirical studies are few, though they address whether farmers are aware of the availability of zinc-biofortified rice in Bangladesh. Glenn Valera et al. (2021) pointed out that the unavailability of seeds, the low yields compared to popular rice varieties, and farmers' unawareness about the health benefits of zinc-biofortified rice were the major barriers to the adoption of this technology in Bangladesh. It is also reported that lower yield (Roy & Eagle, 2015), lack of awareness (Star Business Report, 2023), coarseness of grain, and low market price are also the major reasons for the low adoption of zinc biofortified rice in Bangladesh (Nutrition Connect,

2019). To our knowledge, however, there is no scientific study on what percentage of farmers are aware of the availability of zinc-biofortified rice in Bangladesh.

Using primary data collected from 1301 farmers from seven districts of Bangladesh during January–March 2022, this study examined farmers' awareness about the availability of zinc-biofortified rice and assessed farmers' knowledge of the health benefits of zinc-biofortified rice. The findings of this study can be used to guide the formulation of effective policies to improve the adoption of zinc-biofortified rice in Bangladesh. The chapter is organized as follows: The next section includes the trends in rice production and consumption and the development of zinc-biofortified rice in Bangladesh. Section "Materials and Methods" presents the materials and methods used in this study. Section "Findings and Discussions" presents and discusses the major findings, and the section "Conclusion and Policy Implications" presents the main conclusions and policy implications.

Rice-Based Intervention to Fight Zinc Deficiency in Bangladesh

Rice is the lifeline of Bangladesh's food security (Minten et al., 2013; Mishra et al., 2015; Mottaleb et al., 2017, 2018b; Sayeed & Murshid, 2018; Timsina et al., 2018). It is the major crop and the principal source of livelihood in Bangladesh (BBS, 2023b). In the 2020–21 fiscal year, rice was cultivated on 11.7 million ha of land, which was more than 72% (16.2 million ha) of the total agricultural land of Bangladesh (BBS, 2022). In 2022, 45.33 million people were engaged in agriculture in Bangladesh (BBS, 2023b). It means rice cultivation is the sole source of livelihood for millions of the economically active labor force in Bangladesh.

The temporal changes in rice area, yield, production, consumption, and daily calorie intake from rice are presented in Table 1. It is important to mention here that in terms of yearly per capita consumption (KG) and daily calorie intake from rice,

Table 1 Trends of rice production, consumption and the share of rice in the daily dietary intake in Bangladesh during 1961–2021

	1961	1971	1981	1991	2001	2011	2021
Area (million ha)	8.48	9.30	10.46	10.24	10.66	11.53	11.70
Yield (ton/ha)	1.70	1.60	1.95	2.66	3.40	4.39	4.87
Production (million MT)	9.62	9.94	13.6	18.2	24.2	50.6	56.9
Total domestic supply (million MT)	9.82	11.3	13.9	19.2	25.9	52.6	59.3
Consumption (Kg/capita/year)	171.9	150.5	147.4	159.9	172.9	264.3	257.5
Daily calorie intake (Kcal)	1712.9	1499.4	1468.3	1593.4	1722.5	1756.9	1711.2
Share of rice in daily calorie intake (%)	79.5	73.7	73.5	75.3	73.9	71.9	66.5
Import (million MT)	0.49	0.35	0.08	0.01	0.15	1.29	2.44

Source: Authors based on FAOSTAT (2022a, b)

Bangladesh ranked top in 2021 in the world (Table 1). Currently, Bangladesh is the third-largest rice-producing and fourth-largest rice-consuming country in the world, while the country ranks number one in terms of per capita rice consumption (FAOSTAT, 2022b).

Rice provides 66% of the daily calorie and 52% protein intake per capita in Bangladesh (FAOSTAT, 2022b). However, as rice generally lacks some key vitamins and minerals, the risk of inadequate zinc intake is high among citizens of Bangladesh (Rahman et al., 2016). For example, according to the National Nutrition Micronutrient Survey 2011–12 report (Government of Bangladesh, 2013), 44.6% of children younger than 5 years, 40% of school-age children, and 57.3% of nonlactating and nonpregnant women in Bangladesh were zinc deficient in 2011–12. Zinc is an essential micronutrient that is required for mental and physical development, especially for children. Zinc deficiency is associated with an increasing number of diarrhea, respiratory disease, and stunting cases (Bhutta et al., 1999; Black et al., 2008; Brown et al., 2009; Gupta et al., 2020). Children suffering from zinc deficiency can have long-term cognitive and physical adverse effects, resulting in lower adult wages and a loss of life quality (Schraeder, 1995; UNICEF, 2013; Wieser et al., 2013). According to some studies, zinc deficiency can increase mortality (Walker et al., 2009; Meenakshi et al., 2010), but other studies find no significant effect (Brown et al., 2009; Yakoob et al., 2011). The national loss in GDP due to zinc mineral and vitamin deficiencies in Bangladesh is around US $700 million per year (Dhaka Tribune, 2021).

As rice is the major staple of Bangladesh, experts and scientists stressed that regular consumption of zinc-biofortified rice can meet the daily zinc demand of a person by 60% (Dhaka Tribune, 2023). Thus, to fight zinc deficiency in the most cost-effective way, the government aims to develop and disseminate zinc-biofortified rice in Bangladesh. In 2013, the Bangladesh Rice Research Institute (BRRI) released the first zinc-biofortified rice variety, BRRI dhan62, for aman season rice (Table 2). Since then, nine zinc-biofortified rice varieties have been released in Bangladesh, of which four are suited for the aman season and the remaining five for the boro season. The zinc content in the zinc-biofortified rice ranged from 19 to 27.6 mg/KG

Table 2 Name of the zinc biofortified rice varieties in Bangladesh and year of their release

Name	Year released	Season	Yield (ton/ha)	Zinc content (mg/Kg)	Protein (%)
BRRI dhan62	2013	Aman	3.5–4.5	19	9
BRRI dhan64	2014	Boro	6.0–6.5	24	7.2
BRRI dhan72	2014	Aman	5.7–7.5	22.8	8.9
BRRI dhan74	2014	Boro	7.1–8.3	24.2	8.3
BRRI dhan84	2017	Boro	6.0–8.0	27.6	9.7
BRRI dhan100	2020	Boro	7.7–8.8	25.7	7.8
BRRI dhan102	2022	Boro	8.1–9.6	25.5	7.5
BU dhan2[a]	2016	Aman	4.0–4.5	22	–
BINAdhan-20[a]	2017	Aman	4.48–7.18	26.5	–

Sources: Based on BRRI (2023) and [a]Department of Crop Botany (2023)

(Table 2). In contrast, the average zinc content of ordinary rice grain in Bangladesh is 12.9 mg/kg (Mayer et al., 2010).

Despite the potential benefits, the adoption of zinc-biofortified rice is still low (Nutrition Connect, 2019). To popularize zinc-biofortified rice, HarvestPlus, a research organization under the umbrella of the Consultative Group for International Agricultural Research (CGIAR) and the Department of Agricultural Extension (DAE), distributed seeds, having reached 2,454,000 households in Bangladesh in 2021 (Nuhara, 2021). Despite these efforts, zinc-biofortified rice accounts for less than 1% of the total rice area in Bangladesh (Nutrition Connect, 2019).

It is argued that the lack of awareness across the value chain (farmers, aggregators, millers, wholesalers, and consumers) as well as the lack of awareness among the consumers is the most important barrier to scaling up adoption of zinc-biofortified rice in Bangladesh (Nutrition Connect, 2019). However, studies confirmed that consumers in Bangladesh are willing to pay a premium of 4.6–15.4% for zinc-biofortified rice (Herrington et al., 2023; Imran et al., 2023), which indicates that a market may be available for such a product, but the lack of farmers' adoption limits the development of that market (e.g., Glenn Valera et al., 2021). Using primary data collected from 1301 farmers from seven districts of Bangladesh, this study characterizes the farmers who are aware of the availability and health benefits of consuming zinc-biofortified rice. This study also characterizes farmers who cultivate zinc-biofortified rice.

Materials and Methods

Sampling and Data

The sample of rice farmers used in this study comes from the list of farmers who received zinc-biofortified rice seeds from HarvestPlus. In Bangladesh, HarvestPlus is actively collaborating with the Bangladesh Rice Research Institute (BRRI) in developing and disseminating zinc-biofortified rice and other nutrition-enriched crops. To popularize zinc-biofortified rice, HarvestPlus Bangladesh, in collaboration with the Department of Agricultural Extension (DAE) and BRRI, distributes free sample seeds among farmers (Nuhara, 2021). We communicated personally with the field-level staff of HarvestPlus and collected the list of the farmers who received free zinc-biofortified rice seeds in several districts. After checking the farmers' lists and consulting with HarvestPlus field staff, we conducted surveys in seven districts of Bangladesh during January–February 2022. The sampled districts are: Dinajpur, Faridpur, Gazipur, Habiganj, Jamalpur, Rangpur, and Thakurgaon. The districts are selected in consultation with HarvestPlus, where BRRI is actively working to scale out biofortified rice.

To collect information from the sampled rice farmers, we developed an Open Data Kit (ODK)-based questionnaire suite of tools that allows paperless data collection using Android mobile devices, such as mobile phones and tablets. Thirty

Table 3 Sampled farmers by districts sampled

District	Target sample		Realized sample		
	Zinc rice	Ordinary rice	Zinc rice	Ordinary rice	Sampled total
Dinajpur	100	100	101	99	200
Faridpur	100	100	104	97	201
Gazipur	100	100	0	100	100
Habiganj	100	100	99	101	200
Jamalpur	100	100	97	102	199
Rangpur	100	100	100	100	200
Thakurgaon	100	100	121	80	200
Total	700	700	622	679	1301

Source: Survey 2022

enumerators were recruited to collect data from the sampled farmers. We pre-tested the questionnaire and trained all enumerators before deploying the survey. The survey was approved by the Internal Research Ethics Board (IREB) of CIMMYT (International Maize and Wheat Improvement Center, Mexico). The goal was to interview 200 farmers from each sampled district, of which half would have cultivated zinc-biofortified rice. Thus, we targeted interviewing a total of 1400 rice farmers; half of them would have experience cultivating zinc-biofortified rice. The final sample consisted of 1301 rice farmers, of whom 622 cultivated zinc-biofortified rice and the remaining 679 cultivated conventional (not fortified) rice (Table 3). We could not trace zinc-biofortified rice farmers in the Gazipur district because we had no information about which farmers received zinc-biofortified rice seeds. We selected Gazipur district to examine which rice varieties the farmers of Gazipur district are cultivating. We collected information on farmers' age, sex, education, number of male and female family members, assets, land ownership, and the source of zinc biofortified seeds. To explore farmers' awareness of zinc-biofortified rice and assess their understanding of the benefits of consuming zinc-biofortified rice, we also collected information on whether or not they have heard about zinc-biofortified rice and wheat and what the benefits of consuming zinc-biofortified rice are. In addition, we asked whether or not farmers cultivated zinc-biofortified rice in the 2021–22 boro season and what the source of seeds was.

Conceptual Framework and Estimation Methods

Research Questions

Three seminal works on agricultural adoption laid the foundation of the adoption theory in agricultural economics. Using hybrid maize seeds, Ryan and Gross (1943) developed the first formal model to explain agricultural technology diffusion. The study confirmed that before the actual diffusion of hybrid maize seed, preliminary knowledge on hybrid maize spread quickly, which ultimately contributed to the rapid diffusion of hybrid maize seed in Iowa, USA. The study by Ryan and Gross (1943) concludes that the early adopters of hybrid maize also influenced adoption through demonstration effects on their neighbor farmers.

Griliches (1957) proposed a diffusion theory. He stressed that public-private investment and the economic return from the new technology are the major factors in the rapid diffusion of new agricultural technology. Rogers (1983), in his diffusion of innovation theory, argued that almost all innovation or new technology is associated with some degree of uncertainty. Adopters often seek information about the latest technology from their near peers, especially their subjective evaluation of the new technology. This exchange of information happens through a process of convergence, including interpersonal networks. Based on Rogers (1983), the diffusion of new technology is certainly a social process in which it is necessary to provide evidence and information about the new technology among the members of society for a rapid adoption of the technology. Based on the seminal works reviewed above and following Pannell et al. (2006), we have developed the following working model for adoption (Fig. 1). In this study, we argue that many rice farmers in Bangladesh are unaware of the availability of zinc-biofortified rice and specifically the health benefits of consuming zinc-biofortified rice, which are the fundamental hurdles to the rapid scaling up of zinc-biofortified rice in Bangladesh.

Model Specifications and Estimation Technique

Based on the conceptual framework (Fig. 1), zinc biofortified rice adoption can be modeled as:

$$(Zinc\ rice\ cultivation)_i = f(PKZ_i, AZB_i, HC_i) \qquad (1)$$

Where,

PKZ_i = Prior knowledge or perception of zinc-biofortified rice varieties
AZB_i = Awareness of the benefits of consuming zinc-biofortified rice, and

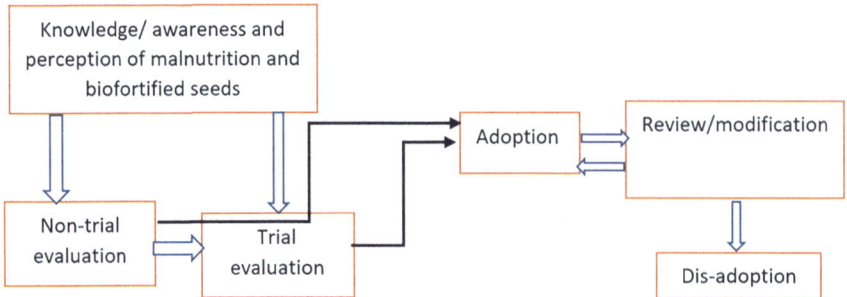

Fig. 1 A schematic flow chart-based working model of adoption

HC_i = Household characteristics, such as level of education and age of the household head and spouse, affiliation to nonfarm sector and NGOs, access to infrastructure, and assets.

The problem is, HC_i can also affect PKZ_i and AZB_i. For example, it is highly likely that more educated farmers are more likely to be aware of the benefits of consuming zinc-biofortified rice. Thus:

$$(PKZ)_i = f(HC_i) \qquad (2)$$

$$(AZB)_i = f(HC_i) \qquad (3)$$

Equations (1), (2), and (3) indicate the need for instrumental variables. However, getting at least two strong instruments that are strongly related $(PKZ)_i$ and $(AZB)_i$ but not related to $(Zinc\ rice\ cultivation)_i$ is difficult. Instead, in this study we estimated the following reduced form equation specified as follows:

$$y_i = \delta_0 + \delta_1 (Age, head)_i + \delta_2 (Years\ of\ schooling, head)_i + \delta_3 (Age, spouse)_i +$$
$$\delta_4 (Years\ of\ schooling, spouse)_i + \delta_4 (Nonfarm\ affiliation)_i +$$
$$\delta_5 (NGO\ membership)_i + \delta_6 (Human\ resource\ index)_i + \delta_7 (Asset\ ownership\ index)_i +$$
$$\delta_8 (Access\ to\ facilities\ index)_i + \delta_9 (Land\ ownership\ index)_i +$$
$$\delta_{10} (Livestock\ ownership\ index)_i + \sum_{5}^{i=1} \varphi_d (DD)_d + \in_i$$

$$(4)$$

Where y_i is a vector of dependent variables, which includes:

- heard about zinc biofortified rice/wheat (y_1, yes = 1, no = 0);
- knows the benefits of the consumption of zinc rice (y_2, yes = 1, no = 0);
- cultivated zinc biofortified rice in *aman* and or *boro* season in 2021–22 (y_3, yes = 1, no = 0);
- purchased or collected seeds (y_4, yes = 1, no = 0) conditional on cultivated zinc biofortified rice (yes = 1); and
- purchased or collected seeds (y_4, yes = 1, no = 0) conditional on cultivated ordinary rice (yes = 1);

In Eq. (4) DD_d is the district dummies for Faridpur (yes = 1), Habiganj (yes = 1), Jamalpur (yes = 1), Rangpur (yes = 1) and Thakurgaon (yes = 1), where Dinajpur and Gazipur districts are set as the base (= 0). In estimating Eq. (4) the logit model estimation process has been applied, specifying the robust standard error.

In Eq. (4), we have included five indices. The indices are calculated by applying the Principal Component Analysis (PCA) method. Annex A shows a detailed description of how these indices are calculated. The human resource index is calculated by combining the number of earners in households, the number of family members who help in agricultural work, the total number of family members, and the number of male family members. In calculating the asset ownership index, we have included the number of mobile phones owned by the household, whether or not they own a bicycle (yes = 1, no = 0), a motorbike (yes = 1, no = 0), a thresher machine (yes = 1, no = 0), a power tiller or two-wheeled tractor (yes = 1, no = 0), irrigation equipment (yes = 1, no = 0), a spray machine (yes = 1, no = 0), a flatbed trailer (yes = 1, no = 0), and a car or rickshaw (yes = 1, no = 0).

The access to facilities index is calculated considering one-way walking time in minutes to the main road and the nearest market. The land ownership index is calculated considering total land cultivated (ha) in 2020–21, whether or not one owns a pond (yes = 1, no = 0), total homestead area (ha), and the size of the largest rice plot (ha). The livestock ownership index is calculated considering the number of cows, goats, buffalo, hens, and ducks owned by the household.

Findings and Discussions

Descriptive Findings

The basic background information of the sampled rice farmers is presented in Table 4. It shows that 570 sampled households (44%) are not aware of the availability of zinc-biofortified rice and wheat, and 731 households (56%) heard about it (Table 4). Based on farmers' awareness of the availability of zinc-biofortified rice and wheat, the background information of the sampled farmers is also presented separately in Tables 4, 5, and 6.

On average, a sampled farmer is 46 years old with 5.5 years of schooling, and the spouse of a sampled farmer is nearly 36 years old with 5.6 years of schooling (Table 4). A sampled household consists of more than five family members, of whom 1.56 were income earners and 1.88 helped with the household's agricultural work. Among the family members, on average, 2.59 are male family members (Table 4). Based on the information on the total number of family members, the number of earners, the number of members who extend help for agricultural work, and the number of male family members, we have calculated a human resource index that ranges from 0 to 1. For a sampled household, a human resource index 1 indicates that the households have very few family members, no earners, no male family members, and no family members to extend help for agricultural work. In contrast, a human resource index of 0 indicates that all the family members of a household are income earners, all of them extend help for agricultural work, and all family members are male. The average human resource index score is 0.72 (Table 4).

A closer scrutiny of Table 4 reveals that households aware of the availability of zinc-biofortified crops are headed by relatively younger and more educated heads and spouses and have more family members who are income earners and agricultural helpers (Table 4). The calculated human resource index is significantly lower

Table 4 Background information of the sampled farmers based whether or not they have heard about biofortified rice and wheat

	All	Heard about zinc biofortified rice and wheat		Mean difference
		No	Yes	
		a	b	(a − b)
No. of observations	1301	570	731	−
Age, household head (years)	46.0 (12.9)	46.5 (13.3)	45.6 (12.5)	0.97* (1.35)
Years of schooling, head	5.5 (4.6)	5.15 (4.59)	5.69 (4.52)	−0.54** (−2.11)
Age, spouse (years)	35.6 (13.0)	36.3 (13.4)	35.0 (12.7)	1.26** (1.74)
Years of schooling, spouse	5.63 (4.23)	5.32 (4.25)	5.88 (4.20)	−0.56** (−2.39)
No. of family members	5.11 (1.88)	5.11 (1.87)	5.12 (1.90)	−0.01 (−0.11)
No. of earners	1.56 (0.83)	1.51 (0.82)	1.60 (0.83)	−0.08** (−1.78)
No. of family members help in agricultural works	1.88 (1.18)	1.81 (1.04)	1.93 (1.27)	−0.11** (−1.75)
No. of male family members	2.59 (1.21)	2.57 (1.25)	2.61 (1.17)	−0.04 (−0.57)
Human resource index	0.72 (0.12)	0.72 (0.11)	0.71 (0.12)	0.01* (1.61)
Households with at least one family member who is a member of an NGO (%)	49.8 (50.0)	41.6 (49.3)	56.2 (49.6)	−14.6*** (−5.29)
Households engaged with nonfarm sector (%)	35.3 (47.8)	31.8 (46.6)	38.0 (48.6)	−6.3*** (−2.35)
Sampled household from Dinajpur (%)	15.4 (36.1)	17.4 (37.9)	13.8 (34.5)	3.6** (1.76)
Sampled household from Gazipur (%)	7.7 (26.6)	17.5 (38.1)	0.00 (0.00)	17.5*** (12.46)
Sampled household from Faridpur (%)	15.4 (36.2)	9.3 (29.1)	20.2 (40.2)	−10.9*** (−5.48)
Sampled household from Habiganj (%)	15.4 (36.1)	16.8 (37.5)	14.2 (35.0)	2.62* (1.30)
Sampled household from Jamalpur (%)	15.3 (36.0)	13.2 (33.8)	17.0 (37.6)	−3.80** (−1.89)
Sampled household from Rangpur (%)	15.4 (36.1)	17.5 (38.1)	13.7 (34.4)	3.86** (1.92)
Sampled household from Thakurgaon (%)	15.4 (36.1)	8.2 (27.5)	21.1 (40.8)	−12.8*** (−6.44)

Note: Standard deviation (first three columns) and t-statistics (last column) are in parentheses
*Significant at the 10% level. **Significant at the 5% level. ***Significant at the 1% level

Table 5 Information on the variables used to calculate indices based on whether or not they have heard about biofortified rice and wheat

	All	Heard about zinc biofortified rice and wheat		Mean difference
		No	Yes	
		a	b	(a − b)
No. of mobile phones owned	1.97	2.00	1.94	0.05
	(1.24)	(1.11)	(1.32)	(0.79)
Own a bicycle (%)	62.0	57.9	65.3	−7.4***
	(48.6)	(49.4)	(47.6)	(−2.72)
Own a motorbike (%)	20.4	17.9	22.4	−4.5**
	(40.3)	(38.4)	(41.7)	(−2.02)
Own a thresher machine (%)	6.9	4.6	8.8	−4.2***
	(25.4)	(20.9)	(28.3)	(−2.97)
Own a tractor/two-wheel power tiller (%)	4.8	2.6	6.4	−3.79***
	(21.3)	(16.0)	(24.5)	(−3.20)
Own an irrigation pump machine (%)	39.2	31.9	44.9	−12.9***
	(48.8)	(46.7)	(49.8)	(−4.78)
Own a spray machine (%)	65.2	61.9	67.7	−5.8**
	(47.7)	(48.6)	(46.8)	(−2.18)
Own a flatbed trailer (%)	1.2	1.2	1.1	0.13
	(10.7)	(11.0)	(10.4)	(0.22)
Own car/auto/rickshaw (%)	10.8	8.4	12.7	−4.30**
	(31.1)	(27.8)	(33.3)	(−2.48)
Calculated asset index	0.81	0.82	0.81	0.008*
	(0.10)	(0.09)	(10.4)	(1.63)
One-way walking distance to the main road from the house (minutes)	12.2	9.89	13.9	−4.05***
	(14.8)	(11.28)	(16.8)	(−4.96)
One-way walking distance to the nearest market from the house (minutes)	19.6	18.5	20.4	−1.90**
	(16.0)	(13.4)	(17.6)	(−2.13)
Calculated access to facilities index	0.84	0.86	0.82	0.04***
	(0.15)	(0.12)	(0.17)	(4.71)
Total agricultural land (ha)	0.57	0.49	0.64	−0.15***
	(0.52)	(0.46)	(0.56)	(−5.27)
Own a pond (%)	43.8	38.9	45.8	−6.9**
	(49.5)	(48.8)	(49.9)	(−2.49)
Total homestead area (ha)	0.07	0.06	0.07	−0.01**
	(0.08)	(0.06)	(0.08)	(−2.40)
Largest rice plot (ha)	0.24	0.26	0.23	0.03***
	(0.18)	(0.20)	(0.17)	(2.69)
Calculated land index	0.88	0.89	0.87	0.03***
	(0.09)	(0.09)	(0.09)	(4.97)
No. of cows own	2.32	2.27	2.36	−0.09
	(2.55)	(2.35)	(2.70)	(−0.62)

(continued)

Table 5 (continued)

	All	Heard about zinc biofortified rice and wheat		Mean difference
		No	Yes	
		a	b	(a − b)
No. of goat own	1.78 (2.44)	1.68 (2.46)	1.86 (2.42)	−0.18* (−1.34)
No. of buffalo own	0.005 (0.09)	0.002 (0.04)	0.007 (0.11)	−0.005 (−1.04)
No. of hen own	12.26 (95.49)	14.1 (124.1)	10.8 (65.0)	3.28 (0.61)
No. of duck own	2.52 (3.55)	2.28 (3.60)	2.71 (3.50)	−0.43** (−2.15)
Calculated livestock ownership index	0.99 (0.05)	0.99 (0.06)	0.99 (0.03)	.001 (−0.48)

Note: Standard deviation (first three columns) and t-statistics (last column) are in parentheses
*Significant at the 10% level. **Significant at the 5% level. ***Significant at the 1% level

Table 6 Knowledge, cultivation and seed source of zinc biofortified rice based on whether or not farmers heard about zinc biofortified rice

	All	Heard about zinc biofortified rice and wheat		Mean difference (t-statistic)
		No	Yes	
		a	b	(a − b)
Know the true benefits of consuming zinc rice (%)	28.9 (45.3)	2.28 (14.9)	49.8 (50.0)	−47.5*** (−21.9)
Cultivated zinc rice (%)	47.8 (49.9)	0.00 (0.00)	85.0 (35.6)	−85.0*** (−56.9)
Purchased or collected seeds in 2020–21 *boro* season (%)	60.3 (48.9)	95.6 (20.5)	32.8 (46.9)	62.8*** (29.8)

***Significant at the 1% level

($p < 0.10$) for households that are aware of the availability of zinc-biofortified rice and wheat. It indicates that households that are aware of the availability of zinc-biofortified rice are headed by a young and educated head and spouse and equipped with a greater number of earners and family members who extend their support to agricultural works.

On average, nearly 50% of the sampled households have at least one family member who is affiliated with an NGO (Table 4, column 2). However, in the case of the households that are aware of the zinc biofortified rice and wheat, more than 56% of them are affiliated to an NGO through at least one member of their family (Table 4). In Bangladesh, the role of the NGO in disseminating seeds and other agricultural technology is widely recognized. For example, maize cultivation has been popularized through NGOs as a tool for poverty alleviation (Mottaleb et al., 2018a). In addition, NGOs in Bangladesh play an important role in the dissemination of hybrid rice (Mottaleb et al., 2015). Zinc-biofortified rice seeds are not generally available in the market but are mainly distributed to farmers for free through NGOs. Thus, NGO affiliation is a crucial determinant of zinc rice familiarity, knowledge of zinc rice, and cultivation of it. In addition, households that are aware of zinc-biofortified rice and wheat are more likely to engage in nonfarm economic activities than others, and the mean difference is negative and statistically highly significant (Table 4).

The number of sampled farm households is drawn almost equally (36%) from the sampled districts, except Gazipur (26%). For our survey, the zinc biofortified rice farmers name were supplied by the HarvestPlus. However, we failed to get any information about who received zinc-biofortified rice seeds from HarvestPlus in Gazipur. Thus, we could not interview any rice farmers in Gazipur who cultivated zinc-biofortified rice.

An examination of asset and livestock ownership and access to important facilities such as the main road and market reveals that farm households that are aware of zinc-biofortified rice and wheat are more likely to have been equipped with bicycles ($p < 0.01$), motorbikes ($p < 0.05$), thresher machines ($p < 0.01$), irrigation machines ($p < 0.01$), and spray machines ($p < 0.05$) (Table 5). While the average calculated asset index for all sampled households is 0.81, it is 0.806 for the households that are aware of zinc-biofortified rice and wheat and 0.82 for other households that are not aware of zinc-biofortified rice and wheat (Table 5). Interestingly, the farm households that are not aware of the zinc-biofortified rice and wheat, on average, reside in closer proximity to the main road and main market than others (Table 5). The findings in Table 5 indicate that farm households located in remote villages and equipped with more agricultural assets are more likely to be informed about zinc-biofortified rice and wheat.

The farm households who are aware of zinc biofortified rice also own more land ($p < 0.01$), are more likely to own a pond ($p < 0.05$), and own a bigger homestead area, but the average size of the largest rice plot is smaller than the farm households who are not aware of zinc biofortified rice (Table 5). The calculated land ownership index for the farm households that did not hear about zinc biofortified rice is 0.89, but it is 0.87 for the sampled farm households that heard about zinc biofortified rice (Table 5). The livestock ownership index is calculated based on the ownership of cows, goats, buffalo, hens, and ducks (Table 5). It shows that the farm households that are aware of the zinc-biofortified rice and wheat own a greater number of goats and ducks than other households (Table 5). The calculated livestock ownership index, however, is almost the same for both types of households, which is 0.99 (Table 5).

Combining the findings of Tables 4 and 5, farm households headed by relatively young and educated head and spouse, and, who are more likely to have been affiliated with NGOs and equipped with more agricultural machinery and livestock, are more likely to hear about zinc-biofortified rice and wheat than other farmers.

Whether the sampled farmers possess true knowledge on the benefits of consuming zinc biofortified rice (yes = 1, no = 0), whether cultivated zinc biofortified rice (yes = 1, no = 0), and the seed source (purchased or collected = 1, received for free = 0) are presented in Table 6. During our interview, we asked farmers to tell us at least one benefit of the consumption of zinc-biofortified rice. Based on the responses of the sampled farmers, we have developed the knowledge variable on zinc biofortified rice, which assumes a value of 1 if the farmer could tell at least one benefit of consuming zinc biofortified rice; otherwise, we assigned a value of 0. It shows that only less than 29% of the sampled farmers know the true benefits of the consumption of zinc biofortified rice (Table 6), while nearly 48% of the sampled farmers cultivated zinc biofortified rice in the 2020–21 boro season, and altogether 60% of the sampled farmers purchased or collected seeds, and the rest received free seeds (Table 6).

A closer scrutiny of Table 6 however shows that the majority of sampled farmers who heard about the zinc biofortified rice, 50% of them know the benefits of consuming zinc biofortified rice, and 85% cultivated zinc biofortified rice in the 2020–21 boro season (Table 6). However, it shows that only 33% of the farmers who heard about zinc-biofortified rice have purchased or collected seeds (Table 6). We found that out of 622 sampled farmers who cultivated zinc rice in the 2020–21 boro season, 479 (77%) of them received zinc biofortified rice seeds for free, while only 143 sampled farmers (23%) purchased or collected seeds.

In Table 7, we have presented the sampled rice farmers' seed sources by district and by whether or not they cultivated zinc-biofortified rice. A total of 679 farmers cultivated ordinary rice, of which 642 farmers purchased or collected their seeds. Only 37 farmers received free seeds (Table 7). In contrast, 622 sampled farmers cultivated zinc-biofortified rice in the 2020–21 boro season, of which 479 received free seeds and 143 farmers purchased or collected seeds (Table 7). Combining

Table 7 Seed source of the sampled rice farmers (number) by whether or not cultivated zinc biofortified rice in 2021–22 boro season

	Dinajpur	Faridpur	Gazipur	Habiganj	Jamalpur	Rangpur	Thakurgaon	Total
Non biofortified rice farmers								
Purchased or collected	97	94	100	99	85	95	72	642
Received free seeds	2	3	0	2	17	5	8	37
Subtotal	99	97	100	101	102	100	80	679
Zinc biofortified rice farmers								
Purchased or collected	24	21	0	28	21	18	31	143
Received free seeds	77	83	0	71	76	82	90	479
Total	101	104	0	99	97	100	121	622
Grand total	200	201	100	200	199	200	201	1301

Source: Survey 2022

findings in Tables 6 and 7 indicate that until now, 10 years after dissemination, the cultivation of zinc-biofortified rice is unsustainable, noncommercial, and mostly dependent on the distribution of free seeds. Based on the findings of our study, this is primarily because, firstly, farmers are not aware of the availability of zinc-biofortified rice and secondly, farmers are not aware of the benefits of the consumption of zinc-biofortified rice.

Econometric Findings

Table 8 presents the marginal effects from the estimated functions applying the logit model estimation process explaining whether (1) the sampled farmers heard about the zinc biofortified rice (yes = 1, no = 0), (2) the sampled farmers possess true knowledge on the consumption of zinc biofortified rice (yes = 1, no = 0), (3) the sampled farmers cultivated zinc biofortified rice in the 2020–21 boro season (yes = 1, no = 0), and (4) the sampled farmers purchased seeds (yes = 1, no = 0). Explaining the estimated coefficients from a logit model is not straight-forward, as the coefficients are the log odds of the probabilities. Thus, econometric findings are mainly based on the calculated marginal effects after logit models.

The years of schooling of the household head positively and statistically significantly correlated in explaining knowledge of the sampled farmers on the health benefits of zinc biofortified rice and wheat ($p < 0.10$) and explaining seed source for ordinary non-biofortified rice ($p < 0.05$) (Table 8). It shows that a one-year increase in the years of schooling of the head of a sampled household on average increases the probability of having proper knowledge on the benefits of consuming zinc-biofortified rice by 1% (Table 8). In addition, a one-year increase in the years of schooling of a sampled head increases the probability of seed purchase in the case of ordinary rice by 0.3% (Table 8). Years of schooling of the head and spouse, however, have no impact on whether farmers cultivated zinc rice or purchased seeds in the case of zinc rice. This is because zinc-biofortified rice cultivation is still dependent on free seed distribution by NGOs in Bangladesh. The highly significant coefficient of the NGO membership variable (yes = 1, no = 0) in explaining whether or not people heard about zinc biofortified rice ($p < 0.01$), whether they knew the health benefits of consuming zinc rice ($p < 0.10$), and whether or not they cultivated zinc rice ($p < 0.01$) indicates that farm households that are affiliated with NGOs are aware of the zinc biofortified rice, and households cultivate zinc biofortified rice only after receiving free seeds from NGOs. This is why there is no single variable that significantly explains the seed source in the case of zinc rice (Table 8). In fact, the Wad Chi2 and the corresponding p-value are insignificant for the equation explaining seed source in the case of zinc rice cultivation (Table 9).

The calculated human resource index, access to infrastructure index, and land ownership index, which range from 0 to 1, are all significant and negative in explaining whether farm households heard about zinc biofortified rice and wheat (yes = 1), whether they are aware of the health benefits of consuming zinc biofortified rice and

Current Status of Zinc-Biofortified Rice Cultivation in Bangladesh 111

Table 8 Marginal effects (dy/dx) after logit model estimation in Table 9 y = Pr (dependent variable) (predict)

	Heard about zinc biofortified rice and wheat (yes = 1)	Knows the health benefits of zinc rice (yes = 1)	Cultivated zinc rice (yes = 1)	Seed purchased or collected (yes = 1) Seed source ordinary rice	
				Zinc rice = 1	Ordinary rice = 1
Age, household head	0.001 (0.00)	−0.0002 (0.00)	−0.0002 (0.002)	0.001 (0.00)	−0.0004 (0.001)
Years of schooling, household head	0.0003 (0.00)	0.01* (0.00)	−0.003 (0.004)	0.01 (0.00)	0.003** (0.002)
Age, spouse	−0.004** (0.00)	−0.002 (0.00)	−0.004* (0.001)	0.001 (0.00)	0.0003 (0.00)
Years of schooling, spouse	0.006 (0.00)	0.004 (0.00)	0.004 (0.004)	−0.01 (0.00)	−0.001 (0.002)
Any member of the household earns from nonfarm sector (yes = 1)	−0.01 (0.03)	0.002 (0.03)	−0.01 (0.03)	−0.11 (0.03)	0.02* (0.01)
Any member of the household is a member of an NGO (yes = 1)	0.10*** (0.03)	0.04* (0.03)	0.08*** (0.03)	0.01 (0.04)	−0.0001 (0.01)
Human resource index	−0.29* (0.15)	−0.31** (0.14)	−0.31** (0.15)	0.20 (0.16)	0.02 (0.05)
Asset index	0.06 (0.18)	0.04 (0.16)	0.07 (0.19)	−0.07 (0.17)	0.05 (0.05)
Access to infrastructure index	−0.65*** (0.14)	−0.38*** (0.12)	−0.42*** (0.14)	−0.14 (0.14)	−0.01 (0.04)
Land ownership index	−0.82*** (0.23)	−0.59*** (0.15)	−0.66*** (0.19)	−0.02 (0.19)	0.13** (0.06)
Animal index	−0.12 (0.23)	0.36 (0.28)	−0.11 (0.24)	1.57 (2.22)	−0.003 (0.04)
District dummies (Dinajpur and Gazipur = 0)					
Faridpur (yes = 1)	0.35*** (0.03)	0.35*** (0.05)	0.17*** (0.05)	−0.01 (0.06)	−0.04 (0.05)
Habiganj (yes = 1)	0.03 (0.09)	−0.02 (0.06)	0.06 (0.06)	0.04 (0.08)	−0.02 (0.04)
Jamalpur (yes = 1)	0.29*** (0.03)	0.38*** (0.06)	0.19*** (0.05)	−0.002 (0.06)	−0.28 (0.13)
Rangpur (yes = 1)	0.17*** (0.04)	0.09 (0.06)	0.17*** (0.05)	−0.03 (0.06)	−0.08 (0.07)

(continued)

Table 8 (continued)

	Heard about zinc biofortified rice and wheat (yes = 1)	Knows the health benefits of zinc rice (yes = 1)	Cultivated zinc rice (yes = 1)	Seed purchased or collected (yes = 1) Seed source ordinary rice	
				Zinc rice = 1	Ordinary rice = 1
Thakurgaon (yes = 1)	0.33*** (0.03)	0.43*** (0.05)	0.22*** (0.05)	0.04 (0.06)	−0.24* (0.13)
Constant					

Note: Number in parentheses are standard errors
*Significant at the 10% level. **Significant at the 5% level. ***Significant at the 1% level

wheat (yes = 1), and whether they have cultivated zinc biofortified rice in the 2020–21 boro season (Table 8). It is important to mention here that all indices are calculated as the highest value minus the estimated value divided by the highest value in the series. It means that households with more land and livestock have smaller indices, and they usually have more family members residing relatively far from the main road and market. Now, the estimated marginal effects show that three indices are negatively associated with three dependent variables (Table 8). It means households with more family members, more agricultural helper family members, residing in relatively remote villages, and owning more agricultural land are more likely to hear about zinc biofortified rice and wheat, more likely to know about the health benefits of consuming zinc biofortified rice, and more likely to cultivate zinc biofortified rice (Table 8).

Again, as we argue that zinc rice cultivation is still mainly driven by free seeds availability from NGOs, the responsible NGOs are aware and distribute seeds only to households that are strictly farm households. That is why the human resource, access to infrastructure, and land ownership indices are highly significant in explaining whether farm households heard about the zinc biofortified rice and wheat (yes = 1), whether they are aware of the health benefits of consuming zinc biofortified rice and wheat (yes = 1), and whether they have cultivated zinc biofortified rice in the 2020–21 boro season (Table 8).

Interestingly, the land ownership index is positive and significant ($p < 0.05$) in explaining the seed source in the case of non-biofortified rice cultivation (Table 8, last column). It indicates that, on average, smallholder farmers are more likely to purchase seeds compared to large farmers (Table 8). Whether this is because smallholder farmers are more market-oriented or because smallholder farmers cannot afford to save their own seeds after meeting demand and are thus forced to purchase seed, this can be an area to investigate further.

The district dummies indicate that compared to Dinajpur and Gazipur districts, which are the base (= 0), the sampled farmers in Faridpur, Jamalpur, Rangpur, and Thakurgaon are more likely to hear about zinc biofortified rice and wheat, know the health benefits of consuming zinc biofortified rice and wheat, and be more likely to cultivate zinc biofortified rice in the 2020–21 boro season (Table 8). For example,

Table 9 Estimation functions applying logit model estimation procedure explaining knowledge on zinc rice, zinc rice cultivation and seed sources

	Heard about zinc biofortified rice and wheat (yes = 1)	Knows the health benefits of zinc rice (yes = 1)	Cultivated zinc rice (yes = 1)	Seed purchased or collected (yes = 1) Seed source ordinary rice	
				Zinc rice = 1	Ordinary rice = 1
Age, household head	0.003 (0.01)	−0.001 (0.01)	−0.001 (0.01)	0.0031 (0.01)	−0.01 (0.02)
Years of schooling, household head	0.0014 (0.02)	0.034* (0.02)	−0.012 (0.02)	0.038 (0.03)	0.12** (0.06)
Age, spouse	−0.02* (0.01)	−0.01 (0.01)	−0.02* (0.01)	0.01 (0.01)	0.01 (0.02)
Years of schooling, spouse	0.03 (0.02)	0.02 (0.02)	0.02 (0.02)	−0.04 (0.03)	−0.02 (0.06)
Any member of the household earns from nonfarm sector (yes = 1)	−0.049 (0.14)	0.01 (0.15)	−0.046 (0.13)	−0.65*** (0.23)	0.69** (0.33)
Any member of the household is a member of an NGO (yes = 1)	0.40*** (0.13)	0.23* (0.14)	0.33*** (0.12)	0.074 (0.21)	−0.0047 (0.40)
Human resource index	−1.17* (0.62)	−1.63** (0.71)	−1.23** (0.61)	1.20 (0.96)	0.88 (1.91)
Asset index	0.26 (0.75)	0.21 (0.83)	0.30 (0.78)	−0.40 (1.02)	1.85 (1.89)
Road index	−2.68*** (0.59)	−1.98*** (0.62)	−1.69*** (0.57)	−0.83 (0.81)	−0.54 (1.63)
Land index	−3.37*** (0.93)	−3.09*** (0.81)	−2.64*** (0.78)	−0.14 (1.09)	4.69*** (1.68)
Animal index	−0.51 (0.93)	1.89 (1.48)	−0.46 (0.97)	9.22 (13.10)	−0.094 (1.37)
District dummies (Dinajpur and Gazipur = 0)					
Faridpur (yes = 1)	1.73*** (0.21)	1.55*** (0.24)	0.71*** (0.20)	−0.088 (0.35)	−0.99 (0.95)
Habiganj (yes = 1)	0.14 (0.24)	−0.13 (0.31)	0.23 (0.23)	0.20 (0.42)	−0.68 (1.04)
Jamalpur (yes = 1)	1.38*** (0.20)	1.71*** (0.25)	0.76*** (0.20)	−0.014 (0.36)	−3.20*** (0.85)
Rangpur (yes = 1)	0.76*** (0.21)	0.43 (0.28)	0.70*** (0.20)	−0.18 (0.38)	−1.62* (0.89)
Thakurgaon (yes = 1)	1.65*** (0.23)	1.91*** (0.25)	0.93*** (0.21)	0.21 (0.34)	−2.86*** (0.89)

(continued)

Table 9 (continued)

	Heard about zinc biofortified rice and wheat (yes = 1)	Knows the health benefits of zinc rice (yes = 1)	Cultivated zinc rice (yes = 1)	Seed purchased or collected (yes = 1) Seed source ordinary rice	
				Zinc rice = 1	Ordinary rice = 1
Constant	5.87***	1.66	4.61***	−10.4	−1.50
	(1.47)	(1.80)	(1.42)	(12.74)	(3.66)
No. of observations	1301	1301	1301	622	679
Wald chi^2 (16)	171.01	166.78	80.12	20.61	68.04
Prob > chi^2	0.00	0.00	0.00	0.1940	0.00
Pseudo R^2	0.17	0.14	0.05	0.03	0.17
Log pseudolikelihood	−787.2	−676.84	−854.4	−324.48	−119.70

Note: Number in parentheses are standard errors
*Significant at the 10% level. **Significant at the 5% level. ***Significant at the 1% level

on average, 35% more farmers in Faridpur district have heard about biofortified rice and wheat and know the health benefits of consuming biofortified rice and wheat ($p < 0.01$), and 17% more farmers cultivated zinc rice in the 2020–21 boro season compared to Dinajpur and Gazipur districts (Table 8). This is probably because HarvestPlus and other NGOs, which are responsible for awareness building and free seed distribution, have worked more in these districts than in Dinajpur and Gazipur districts.

Conclusion and Policy Implications

Bangladesh, with a population of more than 171 million (World Bank, 2022), is highly successful in achieving high agricultural productivity, followed by ensuring food security. Still, hidden hunger, in the form of malnutrition, is a major issue in the country. For example, 28% of children under 5 years of age are stunted, and 9.8% of them are wasted (Development Initiatives, 2020). In addition, 36.7% of women of reproductive age are anemic, and 13% of the total population of the country is undernourished. Studies stressed that 41–44.6% of pre-school-aged children and 57.3% of non-pregnant and non-lactating women in Bangladesh are zinc deficient (Zn).

Rice is the major staple food in Bangladesh. To fight zinc deficiency in the most cost-effective way, the government of Bangladesh has been trying to develop and scale out zinc-biofortified rice and wheat varieties. The first zinc-biofortified rice was released in 2013. Until now, a total of nine zinc-biofortified rice varieties and one zinc-biofortified wheat variety have been released in Bangladesh. However, the uptake of zinc-biofortified rice is very low.

Using primary data collected from 1301 farm households in seven districts of Bangladesh, this study demonstrates that at least 44% of the sampled farmers have not heard about zinc-biofortified rice and wheat, and only less than one-third of the sampled farmers are aware of the true health benefits of consuming zinc-biofortified rice and wheat.

As every new technology and seed in agriculture is associated with risk, prior knowledge and perception of a new technology significantly shape the adoption of a new technology and seeds by farmers in general. Based on the findings, this study argued that the uptake of zinc biofortified rice is low in Bangladesh, and farmers are not aware of the availability and benefits of the zinc biofortified rice varieties. This is why, until now, the cultivation of zinc-biofortified rice in Bangladesh has been unsustainable, which is mainly dependent on the distribution of free seeds.

Based on the findings, to facilitate rapid uptake and consumption of zinc biofortified rice in Bangladesh, this study urges the introduction and strengthening of the provision of information dissemination among farmers on zinc biofortified rice. Specifically, it is important to inform farmers of the dangers of zinc deficiency in the human body and how regular consumption of zinc-biofortified rice can mitigate the problem. In addition, the agronomic performance of the newly developed and released zinc-biofortified rice must be communicated massively among the rice farmers. Resource-poor rice farmers in Bangladesh first think about yield, and only after that may they consider the health benefits of a new rice variety. Thus, it is important to convey the message to farmers that the latest zinc biofortified rice varieties are agronomically superior varieties, the grain quality is good (slender and long), the cooking quality is also good (non-sticky), and the yield is competitive compared.

Field demonstrations in every district during the boro and aman seasons, the organization of farmers' field days, and campaigns and advertisements through public mass communication departments such as radio and television can play important roles in building awareness among the farmers. In addition, international donor agencies can play important roles in raising awareness among farmers in Bangladesh.

Acknowledgement The project grant from the CRP (CGIAR Research Program) WHEAT Agri-Food Systems of the International Maize and Wheat Improvement Center (CIMMYT), Mexico, and the generous financial support of the Asian Development Bank Institute (ADBI), Tokyo, Japan, are greatly acknowledged.

Annexure A: Principal Component Analysis

Principal component analysis (PCA) was used to generate the indices for each household on human resources, asset ownership, access to facilities, land ownership, and livestock ownership based on the information collected from the sampled households. To reduce the dimension of a dataset, PCA is used as a statistical procedure in which a group of variables are aggregated through variables' orthogonal liner combinations. Mathematically, from an initial set of n correlated variables,

PCA creates orthogonal components, where each component is a liner-weighted combination of the initial variables. For n variables:

$PC_1 = a_{11}X_1 + a_{12}X_2 + a_{13}X_3 + \text{---------} + a_{1n}X_n$

"   "

"   "

"   "

$PC_m = a_{1m}X_1 + a_{2m}X_2 + a_{3m}X_3 + \text{---------} + a_{mn}X_n$

where a_{mn} represents the weight for the mth principal component and the nth variable.

The weights for each principal component are given by the eigenvectors of the covariance matrix, as we used the original data. The correlation matrix could be used if the data were standardized. Using the scores generated by the first principal component and the mean and standard deviation of the original data set, the relation indices were computed using the following formula:

$$W_j = \sum_n^i \left[\gamma_i * \left(X_{ij} - \overline{Xi} \right) / \delta_i \right] \quad (7)$$

Where, W_j is the index for each sampled farmers, γ_i represents the weights (scores) assigned to the n indicators on the first principal component; X_{ij} is the original observation of the variables of interest, $\overline{Xi}$ is the mean number of the sampled indicators for the ith farm household of each of the n variables, and δ_i is the standard deviation of each of the variables. Finally, the index is generate for each household as follows:

$$I_n = \frac{\max(W_j) - W_j}{\max(W_j)} \quad (8)$$

In Eq. (8), we have simply subtracted the index value from the highest value in the series and divided it by the highest value. This is done to bring down an index value between 0 and 1.

In generating indices, we have used the following variables:

Index	Variables	Eigenvalue	Proportion
Human resource index	No. of income earners	2.59	0.65
	No. of family members extend help in agricultural works	0.67	0.17
	Household size (no.)	0.49	0.12
	No. of male family member	0.24	0.06

(continued)

Index	Variables	Eigenvalue	Proportion
Asset ownership index	No. of mobile phones owned by the household	1.99	0.22
	Own a bi-cycle (yes = 1)	1.23	0.13
	Own a motorbike (yes = 1)	1.01	0.11
	Own a thresher machine (yes = 1)	0.98	0.10
	Own a two-wheeled power tiller/tractor (yes = 1)	0.89	0.09
	Own an irrigation pump (yes = 1)	0.82	0.09
	Own a sprayer (yes = 1)	0.73	0.08
	Own a flatbed trailer to attach to tiller/tractor (yes = 1)	0.69	0.07
	Own a car/auto rickshaw/van (yes = 1)	0.62	0.06
Access to facilities index	One-way walking distance to the main road from the house (minutes)	1.41	0.71
	One-way walking distance to the nearest market from the house (minutes)	0.59	0.29
Land ownership index	Total land cultivated in 2020–21 season (ha)	1.75	0.44
	Own a pond (yes = 1)	0.99	0.25
	Total homestead area (ha)	0.74	0.18
	The size of the largest rice plot (ha)	0.52	0.12
Livestock ownership index	No. of cows own	1.29	0.26
	No. of goat own	1.02	0.20
	No. of buffalo own	1.00	0.20
	No. of hen own	0.92	0.18
	No. of duck own	0.77	0.15

References

Adesina, A. A., & Baidu-Forson, J. (1995). Farmers' perceptions and adoption of new agricultural technology: Evidence from analysis in Burkina Faso and Guinea, West Africa. *Agricultural Economics, 13*, 1–9. https://doi.org/10.1016/0169-5150(95)01142-8

Adesina, A. A., & Zinnah, M. M. (1993). Technology characteristics, farmers' perceptions and adoption decisions: A Tobit model application in Sierra Leone. *Agricultural Economics, 9*, 297–311. https://doi.org/10.1111/j.1574-0862.1993.tb00276.x

Ara, G., Khanam, M., Rahman, A. S., Islam, Z., Farhad, S., Sanin, K. I., et al. (2019). Effectiveness of micronutrient-fortified rice consumption on anaemia and zinc status among vulnerable women in Bangladesh. *PLoS One, 14*, e0210501. https://doi.org/10.1371/journal.pone.0210501

Bashar, K. (2018). *Moving from Agriculture to food through biofortification*.

BBS. (2022). *Yearbook of agricultural statistics-2021* (33rd Series). Bangladesh Bureau of Statistics (BBS), Statistics and Informatics Division (SID), Ministry of Planning Government, Government of the People's Republic of Bangladesh.

BBS. (2023a). *Agriculture-estimates of major crops*. Available at: http://bbs.gov.bd/site/page/453af260-6aea-4331-b4a5-7b66fe63ba61/. Accessed 15 June 2023.

BBS. (2023b). *Quarterly labor force survey 2022 Bangladesh: Provisional report*. Dhaka. Available at: https://reliefweb.int/report/bangladesh/bangladesh-quarterly-labour-force-survey-2022-provisional-report. Accessed 8 June 2023.

Bhutta, Z. A., Black, R. E., Brown, K. H., Gardner, J. M., Gore, S., Hidayat, A., et al. (1999). Prevention of diarrhea and pneumonia by zinc supplementation in children in developing countries: Pooled analysis of randomized controlled trials. *The Journal of Pediatrics, 135*, 689–697. https://doi.org/10.1016/S0022-3476(99)70086-7

Black, R. E., Allen, L. H., Qar Bhutta, Z. A., Caulfield, L. E., de Onis, M., Ezzati, M., et al. (2008). Maternal and child undernutrition: Global and regional exposures and health consequences. *Lancet, 371*, 243–260. https://doi.org/10.1016/S0140

Brown, K. H., Peerson, J. M., Baker, S. K., & Hess, S. Y. (2009). Preventive zinc supplementation among infants, preschoolers, and older prepubertal children. *Food and Nutrition Bulletin, 30*, S12–S40. https://doi.org/10.1177/15648265090301S103

BRRI. (2023). *Bangladesh rice knowledge bank – Aman rice varieties*. Bangladesh Rice Research Institute (BRRI), Gazipur, Dhaka. Available at: http://knowledgebank-brri.org/brri-rice-varieties/aman-rice-varieties/. Accessed 21 Aug 2023.

Byerlee, D., & Polanco, E. H. (1986). Farmers' stepwise adoption of technological packages: Evidence from the Mexican Altiplano. *American Journal of Agricultural Economics, 68*, 519–527. https://doi.org/10.2307/1241537

Department of Crop Botany, B. S. M. R. A. U. (2023). Digital herbarium of crop plants. *BU dhan2 and BINA dhan20: Main features of these varieties*.

Development Initiatives. (2020). *2020 Global nutrition report: Country nutrition profiles, Bangladesh*. Bristol. Available at: https://globalnutritionreport.org/resources/nutrition-profiles/asia/southern-asia/bangladesh/#overview

Dhaka Tribune. (2021). *Biofortified zinc rice dissemination in Thakurgaon through enrich project*. Available at: https://www.dhakatribune.com/bangladesh/agriculture/251060/biofortified-zinc-rice-dissemination-in-thakurgaon. Accessed 28 Aug 2023.

Dhaka Tribune. (2023). *Bangladesh develops best hi-zinc rice yet*. Available at: https://www.dhakatribune.com/bangladesh/agriculture/238339/bangladesh-develops-best-hi-zinc-rice-yet. Accessed 21 Aug 2023.

Ding, Y., Yu, J., Sun, Y., Nayga, R. M., & Liu, Y. (2023). Gene-edited or genetically modified food? The impacts of risk and ambiguity on Chinese consumers' willingness to pay. *Agricultural Economics, 54*, 414–428. https://doi.org/10.1111/agec.12767

DWHH and IFPRI. (2006). *The challenge of the hunger: Global Hunger Index, facts, determinants and trends*. Deutsche Welthungerhilfe (DWHH), and International Food Policy Research Institute (IFPRI). Available at: https://www.globalhungerindex.org/pdf/en/2006.pdf

FAOSTAT. (2022a). *Food balances (-2013, old methodology and population)*. Rome. Available at: https://www.fao.org/faostat/en/#data/FBSH. Accessed 8 Mar 2022.

FAOSTAT. (2022b). *New food balances* (pp. 1–5). Food and Agriculture Organization of the United Nations. Online database on food balance. Available at: http://www.fao.org/faostat/en/#data/FBS. Accessed 10 Apr 2022.

Feder, G., Just, R. E., Zilberman, D., Development, E., Change, C., & Jan, N. (1985). Adoption of agricultural innovations in developing countries: A survey. *Economic Development and Cultural Change, 33*, 255–298.

Foster, A. D., & Rosenzweig, M. R. (2010). Microeconomics of technology adoption. *Annual Review of Economics, 2*, 395–424. https://doi.org/10.1146/annurev.economics.102308.124433

Galasso, E., & Wagstaff, A. (2019, August). The aggregate income losses from childhood stunting and the returns to a nutrition intervention aimed at reducing stunting. *Economics and Human Biology, 34*, 225–238. https://doi.org/10.1016/j.ehb.2019.01.010. Epub 2019 Mar 18. PMID: 31003858.

Glenn Valera, H., Yamano, T., Pede, V., Puskur, R., Ashraful Habib, M., & Bashar, K. (2021). Impact of nutrition training on long-term adoption of high-zinc rice: A randomized control trial study among female farmers in Bangladesh. In *Virtual: International conference of agricultural economists*. https://doi.org/10.22004/ag.econ.315165

Government of Bangladesh. (2013). *National micronutrient survey 2011–12*. Dhaka. Available at: https://www.gainhealth.org/sites/default/files/publications/documents/bangladesh-national-micronutrient-survey-final-report-2013.pdf. Accessed 28 Aug 2023.

Griliches, Z. (1957). Hybrid corn: An exploration in the economics of technological change. *Econometrica, 25*, 501. https://doi.org/10.2307/1905380

Gupta, S., Brazier, A. K. M., & Lowe, N. M. (2020). Zinc deficiency in low- and middle-income countries: prevalence and approaches for mitigation. *Journal of Human Nutrition and Dietetics, 33*, 624–643. https://doi.org/10.1111/jhn.12791

HarvestPlus. (2017). *Nutritious and hardy: Zinc rice in Bangladesh* (pp. 1–6). CGIAR Research Program on Agriculture for Nutrition and Health (A4NH). Available at: https://www.harvestplus.org/knowledge-market/in-the-news/nutritious-and-hardy-zinc-rice-bangladesh. Accessed 10 Apr 2021.

Herrington, C. L., Maredia, M. K., Ortega, D. L., Taleon, V., Birol, E., Sarkar, M. A. R., et al. (2023). Rural Bangladeshi consumers' (un)willingness to pay for low-milled rice: Implications for zinc biofortification. *Agricultural Economics, 54*, 5–22. https://doi.org/10.1111/agec.12739

Imran, S., Meerza, A., Mottaleb, K., Dsouza, A., Rahaman, M. S., & Sarkar, M. A. R. (2023). Consumers' valuation of a biofortified crop: Evidence from a laboratory experiment. *Agricultural Economics*. https://doi.org/10.1111/agec.12795

Jongstra, R., Hossain, M. M., Galetti, V., Hall, A. G., Holt, R. R., Cercamondi, C. I., et al. (2022). The effect of zinc-biofortified rice on zinc status of Bangladeshi preschool children: A randomized, double-masked, household-based, controlled trial. *The American Journal of Clinical Nutrition, 115*, 724–737. https://doi.org/10.1093/ajcn/nqab379

Mayer, A. B., Latham, M. C., Duxbury, J. M., Hassan, N., Frongillo, E. A., & Biswas, T. (2010). The zinc content of rice in Bangladesh: relationship to soil, production methods, diets and the zinc status of children. *Proceedings of the Nutrition Society, 69*, E334. https://doi.org/10.1017/S0029665110001436

Meenakshi, J. V., Johnson, N. L., Manyong, V. M., DeGroote, H., Javelosa, J., Yanggen, D. R., et al. (2010). How cost-effective is biofortification in combating micronutrient malnutrition? An ex ante assessment. *World Development, 38*, 64–75. https://doi.org/10.1016/j.worlddev.2009.03.014

Minten, B., Murshid, K. A. S., & Reardon, T. (2013). Food quality changes and implications: Evidence from the rice value chain of Bangladesh. *World Development, 42*, 100–113. https://doi.org/10.1016/j.worlddev.2012.06.015

Mishra, A. K., Mottaleb, K. A., Khanal, A. R., & Mohanty, S. (2015). Abiotic stress and its impact on production efficiency: The case of rice farming in Bangladesh. *Agriculture, Ecosystems and Environment, 199*, 146–153. https://doi.org/10.1016/j.agee.2014.09.006

Mottaleb, K. A., Mohanty, S., & Nelson, A. (2015). Factors influencing hybrid rice adoption: A Bangladesh case. *Australian Journal of Agricultural and Resource Economics, 59*, 258–274. https://doi.org/10.1111/1467-8489.12060

Mottaleb, K. A., Rahut, D. B., & Mishra, A. K. (2017). Modeling rice grain-type preferences in Bangladesh. *British Food Journal, 119*, 2049–2061. https://doi.org/10.1108/BFJ-10-2016-0485

Mottaleb, K. A., Kruseman, G., & Erenstein, O. (2018a). Determinants of maize cultivation in a land-scarce rice-based economy: The case of Bangladesh. *Journal of Crop Improvement, 32*, 453–476. https://doi.org/10.1080/15427528.2018.1446375

Mottaleb, K. A., Rahut, D. B., Kruseman, G., & Erenstein, O. (2018b). Changing food consumption of households in developing countries: A Bangladesh case. *Journal of International Food & Agribusiness Marketing, 30*, 156–174. https://doi.org/10.1080/08974438.2017.1402727

Negatu, W., & Parikh, A. (1999). The impact of perception and other factors on the adoption of agricultural technology in the Moret and Jiru Woreda (district) of Ethiopia. *Agricultural Economics, 21*, 205–216. https://doi.org/10.1016/S0169-5150(99)00020-1

Noyon, A. U. (2023). Zinc rice innovation falters as cultivation slumps. *The Business Standard*. Available at: https://www.tbsnews.net/bangladesh/zinc-rice-innovation-falters-cultivation-slumps-588618. Accessed 23 Aug 2023.

Nuhara, S. (2021). Bangladesh welcomes newest zinc rice variety: BRRI dhan100. *HarvestPlus: Zinc rice, News and Insights*. Available at: https://www.harvestplus.org/bangladesh-welcomes-newest-zinc-rice-variety-brri-dhan100/#:~:text=HarvestPlus%20has%20been%20working%20on,2016%2D2020)%3B%20the%20Bangladesh. Accessed 28 Aug 2023.

Nutrition Connect. (2019). Commercialization assessment: Zinc rice in Bangladesh. *Final report for GAIN and HARVESTPLUS*. Available at: https://nutritionconnect.org/sites/default/files/2020-01/191213_Bangladesh_ZincRice_Report_vFINAL.pdf

OECD. (2001). Adoption of technologies for sustainable farming systems, Wageningen workshop proceedings. In *Adoption of technologies for sustainable farming systems* (p. 149). Organization for Economic Co-Operation and Development (OECD).

Pannell, D. J., Marshall, G. R., Barr, N., Curtis, A., Vanclay, F., & Wilkinson, R. (2006). Understanding and promoting adoption of conservation practices by rural landholders. *Australian Journal of Experimental Agriculture, 46*, 1407. https://doi.org/10.1071/EA05037

Rahman, S., Ahmed, T., Rahman, A. S., Alam, N., Ahmed, A. M. S., Ireen, S., et al. (2016). Status of zinc nutrition in Bangladesh: The underlying associations. *Journal of Nutritional Science, 5*, 1–9, e25. https://doi.org/10.1017/jns.2016.17

Rogers, E. M. (1983). *Diffusion of innovations* (3rd ed.). The Free Press, A Division of Macmillan Publishing Co. Inc. citeulike-article-id: 126680.

Roy, S. D., & Eagle, A. (2015). Farmers lose interest in cultivating zinc rice. *The Daily Star*. Available at: https://www.thedailystar.net/country/farmers-lose-interest-cultivating-zinc-rice-171520?fbclid=IwAR2m-PwuYpJUCqREarrcG3cOJG7L1xCrgVasjNE4Pl2oZR0qRQ30y9FceLI. Accessed 23 Aug 2023.

Ryan, B., & Gross, N. C. (1943). The diffusion of hybrid seed corn in two Iowa communities. *Rural Sociology, 8*, 15–24. https://doi.org/citeulike-article-id:1288385

Saltzman, A., Birol, E., Bouis, H. E., Boy, E., De Moura, F. F., Islam, Y., et al. (2013). Biofortification: Progress toward a more nourishing future. *Global Food Security, 2*, 9–17. https://doi.org/10.1016/j.gfs.2012.12.003

Sayeed, K. A., & Murshid, M. Y. (2018). *Rice prices and growth, and poverty reduction in Bangladesh*. Background paper to the UNCTAD-FAO Commodities and Development Report 2017 Commodity Markets, Economic Growth and Development, Rome. Available at: https://www.fao.org/3/I8332EN/i8332en.pdf. Accessed 7 June 2023.

Schraeder, B. D. (1995). Children with disabilities. *Journal of Pediatric Nursing, 10*, 166–172. https://doi.org/10.1016/S0882-5963(05)80079-X

Star Business Report. (2023). Cultivation of zinc-rich rice yet to catch on. *The Daily Star*. Available at: https://www.thedailystar.net/business/economy/news/cultivation-zinc-rich-rice-yet-catch-3252316?fbclid=IwAR2MHDDcgYs6tw1oXQeYvjJvSpMo6uqxGB8R6S0H6TugOEj_FZjPkq8U-4. Accessed 23 Aug 2023.

Sunding, D., & Zilberman, D. (2001). The agricultural innovation process: Research and technology adoption in a changing agricultural sector. In B. L. Gardner & G. C. Rausser (Eds.), *Handbook of agricultural economics* (pp. 207–261). Elsevier B.V. https://doi.org/10.1016/S1574-0072(01)10007-1

Suri, T. (2011). Selection and comparative advantage in technology adoption. *Econometrica, 79*, 159–209. https://doi.org/10.3982/ECTA7769

Timsina, J., Wolf, J., Guilpart, N., van Bussel, L. G. J., Grassini, P., van Wart, J., et al. (2018). Can Bangladesh produce enough cereals to meet future demand? *Agricultural Systems, 163*, 36–44. https://doi.org/10.1016/j.agsy.2016.11.003

UNICEF. (2013). *State of the world's children 2013: Children with disabilities*. United Nations International Children's Emergency Fund (UNICEF). Available at: https://www.unicef.org/media/84886/file/SOWC-2013.pdf. Accessed 6 Nov 2023.

Von Grebmer, K., Bernstein, J., Wiemers, M., Reiner, L., Bachmeier, M., Hanano, A., et al. (2022). *Global Hunger Index: Food systems transformation and local governance*. Bonn/Dublin. Available at: https://www.globalhungerindex.org/pdf/en/2022.pdf. Accessed 21 Aug 2023.

Walker, Fischer C. L., Ezzati, M., & Black, R. E. (2009). Global and regional child mortality and burden of disease attributable to zinc deficiency. *European Journal of Clinical Nutrition, 63*, 591–597. https://doi.org/10.1038/ejcn.2008.9

Waller, B. E., Hoy, C. W., Henderson, J. L., Stinner, B., & Welty, C. (1998). Matching innovations with potential users, a case study of potato IPM practices. *Agriculture, Ecosystems and Environment, 70*, 203–215. https://doi.org/10.1016/S0167-8809(98)00149-2

Wieser, S., Plessow, R., Eichler, K., Malek, O., Capanzana, M. V., Agdeppa, I., et al. (2013). Burden of micronutrient deficiencies by socio-economic strata in children aged 6 months to 5 years in the Philippines. *BMC Public Health, 13*, 1167. https://doi.org/10.1186/1471-2458-13-1167

World Bank. (2022). *World development indicators: Data Bank*. Data Bank, World Development Indicators. Available at: https://databank.worldbank.org/source/world-development-indicators#. Accessed 18 Apr 2022.

Yakoob, M. Y., Theodoratou, E., Jabeen, A., Imdad, A., Eisele, T. P., Ferguson, J., et al. (2011). Preventive zinc supplementation in developing countries: Impact on mortality and morbidity due to diarrhea, pneumonia and malaria. *BMC Public Health, 11*. https://doi.org/10.1186/1471-2458-11-S3-S23

Open Access This chapter is licensed under the terms of the Creative Commons Attribution 4.0 International License (http://creativecommons.org/licenses/by/4.0/), which permits use, sharing, adaptation, distribution and reproduction in any medium or format, as long as you give appropriate credit to the original author(s) and the source, provide a link to the Creative Commons license and indicate if changes were made.

The images or other third party material in this chapter are included in the chapter's Creative Commons license, unless indicated otherwise in a credit line to the material. If material is not included in the chapter's Creative Commons license and your intended use is not permitted by statutory regulation or exceeds the permitted use, you will need to obtain permission directly from the copyright holder.

Biofortified Cereals Increase Dietary Zinc Intake: Wheat and Maize as Case Studies

Swarnim Gupta and Nicola M. Lowe

Introduction

Micronutrient deficiency, also known as hidden hunger, is a major global concern. It compromises immune systems, delays child growth and development and has a debilitating effect on human potential (Bailey et al. 2015; Tulchinsky 2010). Globally, iron and zinc deficiencies are the most widespread mineral micronutrient malnutrition and these often occur concurrently (Sandstead 2000). Recent estimates derived from 24 nationally representative surveys indicate that over half of preschool-aged children and two-thirds of non-pregnant women of reproductive age (WRA) are deficient in at least one core micronutrient (iron, zinc, and folate) (Stevens et al. 2022). Although no region of the world is unaffected from this burden, including high-income countries, the burden is considerably higher in low- and middle-income countries (LMICs). Regionally, South Asia together with sub-Saharan Africa, East Asia and the Pacific is home for three-quarters of preschool-aged children with hidden hunger. Over half (57%) of non-pregnant women of reproductive age with micronutrient deficiencies live in East Asia and the Pacific or South Asia. However, in the UK and USA, it is estimated that 43% and 32% respectively of WRA are deficient in at least one core micronutrient (Stevens et al. 2022).

Hidden hunger is also challenging the agricultural and nutritional research communities because of the ever-rising global population and expanding food demand. The agricultural interventions to boost food quality by improving the nutritive value of edible crops appear to be one of the viable solutions. Biofortification is one of the promising alternatives to alleviate mineral micronutrient deficiency. It involves increasing the nutrient levels in edible plants during the growth period through conventional breeding, mineral fertilization and transgenic approaches either alone or synergistically (Saltzman et al. 2013). While transgenic crops with high nutrient

S. Gupta · N. M. Lowe (✉)
Centre for Global Development, University of Central Lancashire, Preston, UK
e-mail: Sgupta6@uclan.ac.uk; nmlowe@uclan.ac.uk

content, such as golden rice, have not been well received by both consumers and regulatory bodies, programs such as 'HarvestPlus' are exploiting the widely acceptable conventional breeding techniques to enhance the micronutrient content of commonly consumed staples around the world. As a result, biofortified crops like wheat, maize, rice, banana, cassava, potato, capable of assimilating higher concentrations of micronutrients such as zinc, iron and vitamin A have been released in LMICs to benefit a large section of the population who subsist on low cost staple-based diets (Lockyer et al. 2018).

This chapter offers insights not only into the efficacy and effectiveness of studies conducted on zinc biofortified cereals, specifically focusing on wheat and maize, but also their acceptability to producers and consumers which is key to their successful scale-up. Furthermore, the chapter delves into the strengths and weaknesses of this approach to address zinc deficiency, in comparison to supplementation and food fortification, taking the recent contributions and further perspectives from research into consideration.

The Problem of Zinc Deficiency in LMICs

Zinc is indispensable for all biological systems. It is needed for vital functions at cellular and subcellular levels that can be categorized under catalytic, structural, and regulatory roles. Zinc is a component of more than 300 human enzymes and many other proteins and has function in optimal nucleic acid and protein metabolism, cell growth and differentiation, as well cell-mediated immunity (King et al. 2015). Functional consequences of zinc deficiency encompass compromised physical growth, immune capability, reproductive function and neurobehavioural development (King et al. 2015; Caulfield and Black 2004; Brown et al. 2001; Prasad 2013). The impact disproportionally affects settings with low intakes of absorbable zinc resulting in high rates of stunting, increased child morbidity and mortality, and adverse maternal health and pregnancy outcomes. In LMICs, zinc deficiency is responsible for up to 4.4% childhood death and 1.2% of the burden of disease (3.8% in children 6 months to 5 years) (Fischer Walker et al. 2009).

Recent estimates of the prevalence of zinc deficiency among young children and non-pregnant WRA reported it to be >20% for most of the LMICs regardless of the population sub-groups as assessed using most widely used indicator serum/plasma zinc concentrations (PZC) (Gupta et al. 2020). The prevalence of zinc deficiency was as high as 84% in women and 67% in young children in Cameroon and Cambodia, respectively based on population level surveys (Stevens et al. 2022) (Fig. 1). Although data on PZC in men (reported by only four LMICs) were limited, the consistently high prevalence rates, approximately 66% in Malawi, 77% in Kenya, 42.6% in Mexico, and 31% in the Philippines, suggest that zinc deficiency is not confined solely to children and WRA (Gupta et al. 2020).

In LMICs, low dietary diversity coupled with a reliance on low zinc, high phytate foods are the primary contributors to zinc deficiency. Based on inadequate zinc in

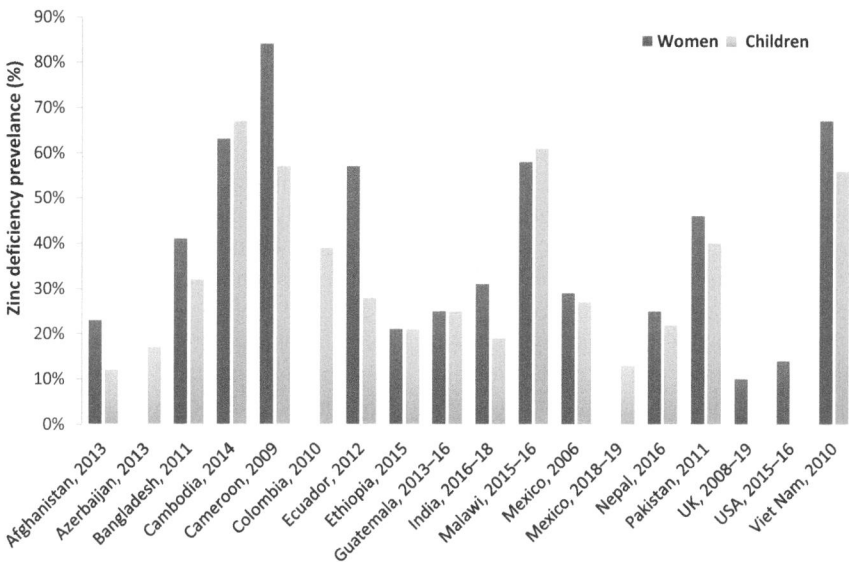

Fig. 1 Prevalence of zinc deficiency among preschool-aged children and women of reproductive age worldwide. Data Source: Stevens et al. (2022). Year indicates the year of the survey

the diet, World Health Organization (WHO) estimates zinc deficiency affects 31% of the global population, with prevalence rates varying from 4% to 73% in different regions (Caulfield and Black 2004). Specifically, prevalence is low (4–7%) in North America and Europe, while it is high in North Africa and the Eastern Mediterranean (25–52%), South and Central America (68%), and in South and Southeast Asia (34–73%). Regardless of the method used to assess zinc status, the situation regarding zinc deficiency in LMICs is a cause for concern.

Zinc Biofortification of Cereals: An Approach with Outstanding Potential for Ameliorating the Problem

Various strategies, such as supplementation, fortification, dietary diversification, and biofortification, have been suggested to improve zinc intake (Gupta et al. 2020; Gibson and Ferguson 1998). The extent of benefits that can be derived from these strategies depends on the context and the resources available for their implementation.

Zinc supplementation in the form of tablets or syrups can rapidly increase zinc intake and address deficiencies in individuals with limited access to diverse diets. It is particularly useful in acute cases of zinc deficiency. However, challenges such as cost, distribution logistics, and long-term compliance can hinder sustained impact and scalability. Despite the known benefits of zinc supplementation on zinc nutriture and health, supplementation in LMICs is primarily limited to being an adjunct therapy for managing diarrhoea in children due to logistical and financial constraints (World Health Organization (WHO) 2006). The coverage of zinc supplementation as an auxiliary therapy for diarrhoea remains low due to insufficient scaling efforts (Black 2019). This approach of therapeutic zinc supplementation is suboptimal for preventing zinc deficiency in children since they can only access supplemental zinc after falling ill, provided their caregivers actively seek diarrhoea treatment (Nasrin et al. 2013). Further, the reachability of any supplementation program to remote or marginalized populations is limited.

Fortification involves adding micronutrients such as zinc to commonly consumed food items such as cereals, flour, or condiments during processing or to food immediately prior to consumption (e.g. multiple micronutrient powders) (Lowe 2021). It can be applied for widespread nutrient deficiency mitigation, either through mass fortification or by targeting vulnerable groups including children and pregnant women. Although fortification has been successful in addressing deficiencies of nutrients such as iodine, zinc fortification faces specific challenges. The bioavailability of fortified zinc can be influenced by food processing and interactions with other dietary components (Lönnerdal 2000). Achieving optimal levels of zinc fortification while maintaining bioavailability can be complex. More importantly, fortification programs require well-established food processing industries and regulatory frameworks which are generally missing in LMICs (Gibson and Ferguson 1998). Food fortification tends to favour urban areas, where the distribution infrastructure for fortified products is more developed compared to rural regions and a possibility of creating a demand for fortification exists due to higher socioeconomic status and greater health literacy levels (Lowe 2021).

Low dietary diversity has been found to be associated with low micronutrient status including zinc (Wiafe et al. 2023). Promoting dietary diversity and or diet modifications encourages individuals to consume a variety of foods that are naturally rich in zinc over a sustained period (Gibson and Anderson 2009). This is highly suited for the needs of LMICs because it does not rely on a constant financial support/infrastructure, which is the case with supplementation and fortification. This approach entails both enhancing zinc intake as well as its absorbability, in contrast to fortification that addresses only intake. While dietary diversity is beneficial for overall health and can provide a comprehensive range of essential nutrients, including zinc, with minimal risk of antagonistic interactions, however, it may be limited by factors such as economic constraints, cultural preferences, and seasonal availability of certain foods. Although promising, the promotion of food-based strategies remains in the early stages of development in LMICs. Programmatic experience with the promotion of home processing techniques to increase absorbable zinc in diet is limited (Brown et al. 2004). Information on locally available, low-cost, culturally acceptable zinc-rich foods and identification of best approach to promote their consumption by those who are at risk of zinc deficiency is required for developing such programmes (Gupta et al. 2020).

Biofortification involves enhancing the nutritional content of in the edible parts crops by increasing the concentration of essential minerals and vitamins. Zinc biofortification specifically focuses on improving the zinc content of mainly staple crops such as rice, wheat, maize, and sorghum, which are major dietary sources for many populations through advanced biotechnology (transgenic) techniques, conventional breeding techniques, agronomic methods (adding zinc fertilizer), or combinations of the latter two (Praharaj et al. 2021).

The transgenic approach, also referred to as genetic modification, involves the insertion of genes necessary for the accumulation of a specific micronutrient in a crop where it would not naturally occur. This technique presents exciting opportunities not only for significantly increasing the nutrient content but also for enhancing its bioavailability. Prominent crops such as rice, wheat, and maize have been genetically modified to enhance their zinc content. Moreover, genetic engineering can also enhance bioavailability by reducing inhibitors or potentially improving the production of enhancers. It has been feasible to increase zinc levels in the edible germ by exploiting this method (Balk et al. 2019). However, transgenic crops face limited acceptance by consumers and regulatory bodies despite their benefits and time-saving advantages over traditional breeding (Kumar et al. 2020; Cui and Shoemaker 2018).

There are several approaches to achieving agronomic biofortification, which include applying zinc fertilizers to soil, leaves, or priming seeds (Praharaj et al. 2021; Bhardwaj et al. 2022). This method is particularly successful in regions where mineral fertilizers are employed, and zinc is added during the manufacturing or distribution process. Importantly, this approach circumvents any limitations posed by low zinc levels in the soil, ensuring optimal zinc accumulation in grains.

Agronomic biofortification with zinc has demonstrated its effectiveness in increasing zinc concentration in crops and offers additional benefits, such as improved yields, even in diverse soil and environmental conditions (Cakmak and Kutman 2018). Furthermore, the utilization of nano-fertilizers for zinc biofortification provides advantages by enhancing the efficiency of micronutrient application, reducing nutrient waste, and minimizing environmental contamination (Dapkekar et al. 2018).

Conventional breeding is a widely utilized method for producing biofortified crop varieties, including those with the capacity to accumulate high amounts of zinc. This process involves crossing parent lines with high nutrient content with recipient lines possessing desirable agronomic traits over multiple generations (Garg et al. 2018). As a result, biofortified crops such as zinc-enriched wheat, rice, and iron-zinc enriched lentils have been successfully developed and released in various countries (HarvestPlus 2023). Table 1 summarizes the zinc-rich cereal varieties that have been released to date.

There are several reasons why zinc biofortification of cereals holds exceptional promise in tackling zinc deficiency particularly in a resource limited setting:

Accessibility and Affordability: Cereal crops, especially wheat, rice, and maize, are widely consumed by large populations, making them an ideal vehicle for

Table 1 Zinc-rich cereal varieties released across the world

Country	Crop	Variety released
Bangladesh	Wheat	BARI Gom -33
	Rice	BRRI Dhan62, BRRI Dhan64, BRRI Dhan72, BRRI Dhan74, BU Aromatic Hybrid Dhan-1, Binadhan 20, BU Aromatic Dhan-2, BRRI Dhan84, BRRI dhan100, BRRI dhan102
Bolivia	Wheat	INIAF Okinawa
	Rice	CIAT BIO-44 +Zinc
Brazil	Wheat	BRS 331
Colombia	Rice	Fedearroz BIOZn 035
	Maize	BIO-MZn01, SGBIOH2, SGBIOH6
El Salvador	Maize	CENTA Porrillo 2020
	Rice	CENTA A-Nutremas
India	Wheat	BHU-3, Zn-Shakti, BHU-1, BHU-5, WB-02, HPBW01, BHU-25, BHU-31, HUW 711, HI 8777 (DURUM), MACS 4028 (DURUM), PBW 757, HI 1633, PBW 771, MACS 4058, DBW 332
	Rice	DRR Dhan 49
	Sorghum	Parbhani Shakti
Indonesia	Rice	INPARI IR Nutri Zinc, Inpara 11 Siam HiZInc, Inpara 12 Mayas
Mexico	Wheat	Nohely F2018
Nepal	Wheat	Zinc Gahun1, Himgange, Panchakoshi, Zinc Gahun2, Zinc wheat 3, Borlaug 100
Nicaragua	Rice	INTA Las Minas
	Maize	Fortinica, INTA-Nutremas
Pakistan	Wheat	Zincol-2016, Akabar-2019, Nawab-21, TARNAB-REHBAR, TARNAB-GANDUM-I
Honduras	Maize	DICTA B02, DICTA B03
Guatemala	Maize	ICTA HB-18ACP+Zn, ICTA B-15ACP+Zn, Fortaleza 17

Data Source: Harvest Plus webpage for Database of Biofortified Crops Released (HarvestPlus 2022)

delivering increased zinc intake to vulnerable communities. Biofortified crops can be easily integrated into existing agricultural practices and local food systems, ensuring accessibility and affordability for the target populations.

Non-disruption of Usual Dietary Behaviours: In contrast to dietary diversification approaches, biofortification generally requires no change in consumer behaviour because it has minimal impact on the sensory attributes of the crops involved. This aspect enhances the acceptability and sustainability of biofortification as an effective intervention for combating zinc deficiency.

Sustained Impact: Once biofortified varieties are introduced and adopted into mainstream seed markets, they can consistently contribute to combatting zinc deficiency. After the successful development of the biofortified plant, its seeds can be widely distributed and continually cultivated by farmers year after year. Following the initial investment in the breeding program, ongoing costs are minimal, although support may be necessary to ensure optimal fertilizer application to maximize the crop's zinc content in regions where soil zinc levels are low. Unlike interventions relying on supplementation or fortified foods, which may encounter challenges with long-term implementation and compliance, biofortified crops offer a continuous source of zinc in the regular diet.

Nutrient Synergy: Cereals are often consumed alongside other staple foods, including legumes, vegetables, and animal products. By increasing the zinc content in cereals, the overall dietary zinc intake can be improved, as these foods are often consumed together, leading to a synergistic effect on nutritional status. Attempts to biofortify wheat have also demonstrated an increase not only in zinc content but also in minerals such as iron and selenium (Lowe et al. 2020; Gupta et al. 2022a).

Improved Agronomic Traits: Biofortification programs also consider agronomic traits, such as yield, disease resistance, and climate resilience, in addition to nutritional enhancement. This ensures that the biofortified crop varieties are not only high in zinc but also perform well in terms of productivity and resilience to environmental stresses, benefiting farmers and encouraging wider adoption.

Comprehensive Approach: Biofortification enables the targeted delivery of nutrients to populations at risk of zinc deficiency, including communities with limited access to diverse diets and those heavily reliant on staple cereals within their local grain production, processing, and consumption systems. This strategy also extends its coverage to populations who may be difficult to reach through supplementation programs. Moreover, it addresses nutritional gaps within specific vulnerable groups, such as WRA and young children, who are particularly susceptible to zinc deficiency. In cases where the equitable household distribution of other zinc-rich foods may not be feasible, this approach remains crucial. Recent data have also highlighted that even adolescent and adult males in LMICs may also be zinc deficient (Gupta et al. 2020). Therefore, this approach bridges the gap between the typical dietary intake and the recommended levels for all population groups without raising concerns about excess consumption.

Analysing Strategies to Improve Zinc Intakes: A Fortification and Biofortification Case in Pakistan

Supplementation is effective for targeted interventions but is impractical for widespread use due to cost and distribution challenges. Achieving dietary diversification remains a distant and complex long-term goal, necessitating a major transformation of agricultural and food systems and the reduction of global inequalities through international political commitment and commercial incentives. In such a situation, only food fortification and biofortification provide practical solutions to address hidden hunger at the population level in the medium term to long term.

In Pakistan, micronutrient deficiencies impact urban and rural populations spanning all geographic regions and income brackets, hence there is a need for a comprehensive, population-wide strategy to effectively address this widespread challenge (Government of Pakistan and United Nations Children's Fund (UNICEF) 2023). It is worthwhile to explore the strengths and weaknesses of these two approaches, especially in the context of Pakistan, where fortification and biofortification initiatives were independently initiated but implemented concurrently in 2016. The Food Fortification Program (FFP), supported by UK Aid, aimed to address vitamin A, iron, and zinc deficiencies, particularly among women and children. An independent consortium evaluated the FFP after 5 years (e-Pact Consortium 2021). Concurrently, Zincol-2016, a biofortified wheat variety was released in collaboration with the HarvestPlus program and the Pakistan National Agriculture Research Centre. A biofortification project, BiZiFED research program, funded by UKRI Global Challenges Research Fund, was initiated to generate data on the effectiveness, acceptability and feasibility of Zincol-2016 (Lowe et al. 2020, 2018; Ohly et al. 2019).

In 1965, Pakistan introduced mandatory fortification of oil and ghee with Vitamin A, but by 2011, national survey revealed persistently high deficiencies in pregnant women (Bhutta et al. 2011). In 2016, a 5-year FFP was launched to combat deficiencies in vitamin A, iron, and zinc, focusing on women and children, and expected to reach 150 million people (e-Pact Consortium 2019). To address specific deficiencies, the FFP utilized different vehicles for fortification. Vitamin A and D were added to oil and ghee, which are commonly used in meal preparation and had the potential for broad distribution, promoting equity. For iron, folate, and zinc, the program turned to fortification of wheat flour, a staple used in various forms of bread consumed daily throughout the year. The approach revolved around enhancing the availability of fortified food items, generating consumer interest, and establishing a favourable setting for food fortification. This all-encompassing strategy entailed providing technical support to local and provincial government agencies, forming partnerships with industry stakeholders, and advocating for the cause to both policymakers and the public.

The final evaluation of the oil and ghee fortification program showed significant progress in enhancing fortification standards and increasing the number of registered mills. This led to the mills achieving adequate levels of vitamins A and D and being on track to meet the annual production target of 2.5 million metric tonnes of fortified oil/ghee by 2021 (e-Pact Consortium 2021). Unlike oil and ghee, the fortification of wheat flour with iron, folate, and zinc was carried out by commercial flour mills on a voluntary basis. In 2020, the FFP faced COVID-related challenges and a wheat shortage affecting prices. Despite this, it improved premix access and micro feeder installation, enabling more mills to fortify in the future if incentives align. However, it fell short of its goal to provide 1.5 million metric tons of fortified wheat flour annually, with fortified flour comprising less than half of the mills' total production (e-Pact Consortium 2021). This approach required a strong public demand for fortified products to incentivize millers and retailers to adopt the fortification strategy and increase supply. The government's control over wheat prices in Pakistan did not offer a compelling economic incentive for voluntary fortification, as producers cannot pass the cost on to consumers. Further, this strategy raises concerns about equity, as the poorest individuals might not have access to these premium products. A potential solution could involve government subsidies to offset the cost of premix and fortification, although this might conflict with the original sustainability goal of fortified wheat flour production. Alternatively, the introduction of legislation for mandatory flour fortification could be considered, although this process is intricate and time-consuming in Pakistan due to the decentralization of decision-making to provincial governments. Additionally, the choice to focus on large commercial roller mills posed a challenge to the flour fortification program's scalability. This is because the mills process only a portion of the wheat flour consumed in Pakistan, estimated to be between 40 to 60% of household wheat flour procurement (Ansari et al. 2018). The rest of the household flour comes from wheat grain kept for self-consumption by farmers or received as in-kind payment by farm laborers, which is milled in numerous small local mills called "chakkis" found in both urban and rural areas across the country.

In 2016, Pakistan introduced Zincol-2016, its first zinc biofortified wheat variety. The BiZiFED program, initiated in 2017, assessed the viability of using this biofortified wheat to combat zinc deficiency on a population scale. The program comprised an efficacy trial from 2017 to 2019 (Lowe et al. 2018) and an effectiveness trial from 2019 to 2021 (Lowe et al. 2020), aiming to study health outcomes in women, adolescent girls, and children, assess crop performance under various conditions, and identify barriers and enablers for scaling up adoption. The study showed that Zincol-2016 had a significantly higher zinc content compared to the Galaxy control, resulting in an increased daily zinc intake for participants (Gupta et al. 2022a; Lowe et al. 2021). Even when local farmers grew Zincol-2016 under real-world conditions with some technical support for zinc fertilizer application, the grain maintained satisfactory zinc levels (Gupta et al. 2022a). Importantly, it was found that the enhanced zinc content in Zincol-2016 did not lead to higher phytate levels, which meant that the bioavailability of zinc was comparable and had the potential for improved absorption compared to the non-biofortified variety (unpublished). Although the study did not demonstrate significant increments in height/weight based anthropometric measurements in adolescent girls and young children after consuming biofortified wheat for six months, there were signs of improved health outcomes related to upper respiratory tract infections toward the end of the intervention period (Gupta et al. 2022b; Gupta et al. 2023) as well as a modest increase in head circumference in children, favouring the biofortified group (unpublished). Notably, existing biomarkers lack the sensitivity to detect subtle changes in dietary zinc intake although one novel biomarker appeared to have captured this. Further details regarding the randomized controlled trials' findings are elaborated in the subsequent sections. Consumer acceptability of biofortified flour discussions revealed that community members and elders appreciated the potential health benefits of biofortified flour (Mahboob et al. 2022). Despite concerns about potentially higher prices compared to standard flour, there was a willingness among consumers to pay a bit extra for the health benefits it offered. Numerous farmers opted to persist in cultivating Zincol-2016, due to its favourable yield and nutritional benefits, as elaborated in section "Acceptability of Zinc-Biofortified Cereals: Consumer Perception and Regulatory Considerations" of this chapter. Nonetheless, they conveyed a requirement for government subsidies to help mitigate the expenses associated with fertilizers needed to enhance the zinc content in wheat grain (Ceballos-Rasgado et al. 2022).

Overall, establishing a supportive environment through policies and programs is crucial for scaling up both wheat flour fortification and biofortification in Pakistan. Both approaches rely on collaborations within the food value chain. However, when considering scalability, biofortification shows promise. According to the latest HarvestPlus report from 2022, the market share of zinc biofortified wheat in Pakistan is expected to reach 20% of the certified seed sector in 2022, benefiting over 1.4 million households growing these varieties (HarvestPlus 2022).

Zinc Content in Traditional vs. Biofortified Crops: What the Data Shows?

The world's primary cereal crops, including maize, rice, and wheat, are cultivated across extensive areas globally. Their combined annual yield, reaching approximately 2.8 billion tons of grain according to Food and Agriculture Organization Statistical Database (FAOSTAT), highlights their paramount significance (Food and Agriculture Organization of United Nation 2023). These crops play an integral role in diets, societies, and economies worldwide, especially in densely populated developing regions. The global demand for all three cereals is steadily increasing, a trend expected to persist until the middle of this century. Consequently, these major cereals offer pivotal opportunities for improving nutritional outcomes.

Wheat

Globally, about 20% of calories come from wheat. In some countries, it is more than 70%. Thus, increasing zinc levels in wheat grain could deliver more zinc to people whose diet relies directly or indirectly on wheat-based food and could help mitigate zinc deficiency. In Asian and African countries, along with international organizations such as International Maize and Wheat Improvement Center (CIMMYT), International Center for Agricultural Research in the Dry Areas (ICARDA), HarvestPlus, are actively developing bio-fortified wheat varieties (Wani et al. 2022). Their collaboration with national research institutes has resulted in the successful development of zinc rich wheat varieties in countries including India, Pakistan,

Nepal, Bangladesh, and others Latin American countries, as outlined in Table 1. Zinc biofortification of wheat has already gained momentum in India with at least 16 high zinc varieties released over the past 7 years.

Pakistan has also introduced five zinc-biofortified wheat varieties since 2016 (Table 1).

The year 2020 marked a significant moment in the release of biofortified wheat varieties. A collaboration between CIMMYT and the Nepal Agricultural Research Council (NARC) resulted in a notable achievement—the introduction of six new wheat varieties in Nepal during that year (see Table 1). Five of these varieties were derived from crosses with wild relatives and contained 20–40% more zinc and iron content compared to local crops. These new varieties not only excelled in terms of yield but also demonstrated enhanced disease resistance in comparison to existing types.

The zinc content of biofortified wheat varieties released worldwide is summarized in Table 2. On an aggregate level, biofortified wheat contains approximately 50% more zinc as compared with non-biofortified varieties.

Maize

Increasing zinc levels in maize grain holds the potential to provide greater zinc intake to people whose diets rely directly or indirectly on maize-derived foods, offering a promising solution to mitigate zinc deficiency, particularly in Africa and South America. Guatemala has taken the lead by releasing zinc-enhanced maize hybrids. Notably, the ICTA HB-18 variety has a 15% higher zinc content compared to other commercially available varieties. Additionally, tortillas produced from ICTA B-15 exhibit an increase in zinc content of up to 60% compared to tortillas made from other commercial varieties (Maqbool and Beshir 2019).

In addition to Guatemala, the CIMMYT has achieved notable success in the development and introduction of zinc-enriched maize varieties in countries including Honduras, Colombia, Nicaragua. According to the HarvestPlus database, a total of 11 high-zinc maize varieties have been launched thus far, containing an additional 13 mg per kg of zinc compared to non-biofortified varieties (Table 2).

Other Cereals

Rice is the world's most vital crop, with over half of the global population heavily reliant on it for sustenance. This dependency makes high zinc rice varieties a crucial intervention in combating zinc deficiency, particularly in regions where daily rice consumption is prevalent. Remarkably, more than half of the 18 high zinc rice lines released thus far were introduced in Bangladesh, a country known for its high per capita rice consumption (144.5 kg/year) (Saha et al. 2021). What is even more

Table 2 Zinc content of biofortified cereal varieties released worldwide

Biofortified crop	Variety name	Release year	Zinc content (ppm)	Comparison with traditional varieties (ppm)
Zinc wheat	Nohely F2018	2018	31	37 vs. 25
	BRS 331	2012	37.3	
	INIAF Okinawa	2019	32	
	Zincol-2016	2016	37	
	Akhbar-2019	2020	37	
	Nawab-21	2021	37	
	TARNAB-REHBAR	2023	34	
	TARNAB-GANDUM-I	2023	36	
	Zinc Gahun-1	2020	38	
	Himgange	2020	54	
	Panchakoshi	2020	39.4	
	Zinc Gahun-2	2020	39.4	
	Zinc wheat-3	2020	48	
	Borlaug100	2020	31	
	BHU-3	2014	30.5	
	Zn-Shakti	2014	34.2	
	BHU-1	2013	34.8	
	BHU-5	2013	29.5	
	WB-02	2017	31	
	HPBW01	2017	31	
	BHU-25	2018	31	
	BHU-31	2018	39.5	
	HUW 711	2019	31	
	HI 8777 (DURUM)	2018	43.6	
	MACS 4028 (DURUM)	2018	40.3	
	PBW 757	2018	42.3	
	HI 1633	2020	41.1	
	PBW 771	2020	41.4	
	MACS 4058	2020	37.8	
	DBW 332	2021	40.6	
	BARI-Gom33	2017	33	
Zinc maize	Fortinica	2018	34.9	33 vs. 20[a]
	INTA-Nutremas	2018	35	
	DICTA B02	2017	34.5	
	DICTA B03	2017	35.1	
	ICTA HB-18ACP+Zn	2018	31	
	ICTA B-15ACP+Zn	2018	30	
	Fortaleza 17	2020	32	
	CENTA Porrillo 2020	2020	32	
	BIO-MZn01	2018	34.5	
	SGBIOH2	2019	33	
	SGBIOH6	2020	32	

(continued)

Table 2 (continued)

Biofortified crop	Variety name	Release year	Zinc content (ppm)	Comparison with traditional varieties (ppm)
Zinc rice	INTA Las Minas	2020	22	25 vs. 16
	CENTA A-Nutremas	2019	22.8	
	Fedearroz BIOZn 035	2021	26	
	CIAT BIO-44 +Zinc	2019	22	
	INPARI IR Nutri Zinc	2018	25	
	Inpara 11 Siam HiZInc	2022	33.9	
	Inpara 12 Mayas	2022	29.8	
	DRR Dhan 49	2018	25.2	
	BRRI Dhan62	2013	20	
	BRRI Dhan64	2014	24	
	BRRI Dhan72	2015	23	
	BRRI Dhan74	2015	24.2	
	BU Aromatic Hybrid Dhan-1	2016	22	
	Binadhan 20	2017	27.5	
	BU Aromatic Dhan-2	2016	22	
	BRRI Dhan84	2017	27.6	
	BRRI dhan100	2021	25	
	BRRI dhan102	2022	25.5	

Data Source: HarvestPlus Database of Biofortified Crops Released (HarvestPlus 2022)
[a]Zinc content for traditional variety is adopted from Prasanna et al. (2019)

remarkable is that these releases (all except two) occurred within a short span of just four years, highlighting the concentrated efforts towards addressing zinc deficiency in the region. Overall, high zinc rice can provide around 50% more zinc as compared to traditional rice but some varieties such as INPARA developed and released in Indonesia can contain up to twice as much as zinc in the non-biofortified rice (Sitaresmi et al. 2023). Since 2019, Latin American countries including Nicaragua, El Salvador, Colombia, Plurinational State of Bolivia have also released their first biofortified zinc rice.

In 2018, International Crops Research Institute for the Semi-Arid Tropics (ICRISAT) launched India's inaugural biofortified sorghum variety, 'Parbhani Shakti' (ICSR 14001), distinguished by its elevated iron and zinc levels compared to regular sorghum. Subsequent efforts have been directed towards expanding its production and dissemination in the sorghum-consuming regions of central India. This recently introduced variant has an average grain concentration of 45 ppm (parts per million) of iron and 32 ppm of zinc, surpassing conventional varieties that typically contain 30 ppm of iron and 20 ppm of zinc, respectively (Gaikwad et al. 2020; Kumar et al. 2018). Notably, it also offers a higher protein content at 11.9%, compared to the typical 10% found in most sorghum types, and a lower phytate content (4.14 mg/100 g) as opposed to the usual 7.0 mg/100 g, thereby enhancing nutrient absorption.

Acceptability of Zinc-Biofortified Cereals: Consumer Perception and Regulatory Considerations

Biofortified cereals enriched with essential micronutrients such as zinc have garnered increasing attention as potential solutions to malnutrition and nutrient deficiencies, particularly in regions where staple foods constitute a major part of the diet. However, their success hinges not only on scientific efficacy but also on the acceptance of these biofortified varieties by consumers and the endorsement of regulatory bodies (Bouis and Saltzman 2017). The earliest and most prominent biofortified crop, Golden Rice, created in 2000 is an example (Dubock 2017). Despite the open availability of licensing for Golden Rice (transgenic) and its derivatives, along with firmly documented nutritional benefits, their on-farm utilization has been impeded until now. This hindrance is due to cautiousness surrounding public health and environmental issues, coupled with substantial adverse publicity from anti-biotechnology interest groups (Listman et al. 2019). In this section, we will focus on the consumer acceptability of biofortified products resulting from selective breeding and/or mineral application, which are generally more acceptable compared to the transgenic approach.

Consumer acceptance of zinc-biofortified cereals is a crucial determinant of their viability. While enhancing the nutritional content of staple crops is an endeavour with far-reaching benefits, it's essential to gauge how these modified varieties resonate with local preferences and cultural norms. Factors such as taste, texture, appearance, and cooking methods can significantly influence consumer adoption. Several studies have provided insights into this aspect. Sensory evaluations, focus groups discussions (FGDs), and surveys have been conducted to assess the palatability and sensory qualities of zinc-biofortified cereals (Woods et al. 2020; Rizwan et al. 2021; Gannon et al. 2019; Mahboob et al. 2020, 2022; Talsma et al. 2017). These studies have shown that while consumers prioritize taste, they are often willing to embrace the health benefits of biofortified options if the changes in taste and appearance are minimal.

A mixed-methods study was conducted alongside a cluster-randomized controlled effectiveness (BiZiFED2) trial in the Peshawar region, Pakistan from November 2020 to July 2021. This study involved semi-structured FGDs with farmers who grew Zincol-2016 wheat for the trial. Additionally, a year after the study was completed, a survey was conducted with 686 farmers in Punjab province, Pakistan's main wheat-growing region, to ascertain if they had grown biofortified Zincol-2016 variety again in the subsequent season. The findings revealed that 47% of participants continued cultivating Zincol-2016 wheat after the trial had ended. Motivations included seed availability, high grain yields, disease resistance, improved flour quality, and nutritional benefits. Farmers appreciated the flour taste and texture and consumed it at home. Qualitative analysis from focus groups identified that technical and financial support, better grain quality, and health advantages promoted scaling up, while challenges encompassed unfamiliarity with biofortification, production costs, and external threats such as COVID-19 pandemic (Ceballos-Rasgado et al. 2022).

In order to mainstream zinc-biofortified cereals into food systems and dietary practices, it is imperative to systematically consider consumer preferences, cultural intricacies, and regulatory demands. Equally vital is the provision of essential resources and training for cultivators. This endeavour necessitates continuous collaboration among researchers, policymakers, food industries, and communities, encompassing both producers and consumers, as they collectively address the multifaceted challenges in this pursuit.

Biofortified Cereals Increase Zinc Intake: Evidence Available from Efficacy and Effectiveness Trials

Several studies have shown that the zinc content of staple crops can be enhanced through conventional breeding or the application of minerals and can lead to several other desirable traits (Lockyer et al. 2018; Cakmak and Kutman 2018; Rashid et al. 2019; Nestel et al. 2006). However, there has been limited research to confirm the translation of this increase in zinc content to benefits for human health. Nonetheless, these studies do indicate a successful incremental increase in zinc intake when consuming biofortified cereals over non-biofortified cultivars.

The studies conducted on biofortified cereals, including modest-scale investigations into the efficacy of conventionally bred biofortified cereals, revealed a distinct increase in zinc intake ranging from 21% to 169% over the control cereal (non-biofortified), depending on the population subgroup. Out of the nine studies listed in Table 3, six tested the efficacy of zinc biofortified wheat (Gupta et al. 2022a; Lowe et al. 2021; Sazawal et al. 2018; Rosado et al. 2009; Signorell et al. 2019, 2023), while one each tested maize (Chomba et al. 2015) and rice (Jongstra et al. 2022). Only one study explored the usefulness of high-iron and high-zinc millet among Indian children (<2 years old) (Mehta et al. 2022). Consuming biofortified pearl millet provided 1.5 mg of daily zinc, compared to the 0.5 mg provided by the control. This scrutiny indicates that when included as a dietary cornerstone, children receive nearly 40% of their zinc requirements from biofortified pearl millet alone. Despite this increased zinc intake over nine months, high-iron and high-zinc pearl millet did not significantly improve zinc biomarkers or growth compared to the control.

Although a bio-efficacy study specifically for maize has not been conducted at the time of writing, the absorption of zinc from consuming high-zinc maize was investigated by Chomba and co-workers (Chomba et al. 2015). Their study demonstrated that the total daily zinc intake from biofortified maize (5.0 mg) was significantly higher ($P < 0.001$) than that from the control maize (2.3 mg) among young rural Zambian children. While the group found no significant difference in the fractional absorption of zinc between the control maize (0.28 mg) and the biofortified maize (0.22 mg), the daily absorption of zinc from the biofortified maize (1.1 mg) was higher ($P < 0.001$) than that from the control maize (0.6 mg). This is because the net absorbed zinc is a function of both fractional zinc absorption and the total

Table 3 Studies assessing the impact of consuming zinc biofortified cereals and millets

	Study	Country	Type	Biofortification	Population	N	Additional daily zinc intake
Wheat	Gupta et al. (2022a)	Pakistan	Effectiveness	Conventional breeding (Zincol-2016) + agronomic	Adolescent girls aged 10–16 years	517	21%
	Lowe et al. (2021)	Pakistan	Efficacy	Conventionally bread (Zincol-2016) + agronomic	NPNL women of reproductive age (16–49 years)	50	30–60%[a]
	Sazawal et al. (2018)	India	Unclear	Conventionally bread (PBW 550) + agronomic	Children aged 4–6 years	6050	50%
	Sazawal et al. (2018)	India	Unclear	Conventionally bread (PBW 550) + agronomic	NPNL woman of child-bearing age (15–49 years)	6050	39%
	Rosado et al. (2009)	Mexico	Efficacy (absorption)	Conventionally bread (combined from six landraces and included: DGO95.1.17; DGO95.3.2; CHIH95.2.1; CHIH95.2.47; CHIH95.3.47; JAL95.4.10 and LGP2 and LGP12)	NPNL adult women aged 18–42 years	27	68–72%[a]
	Signorell et al. (2019)	Switzerland	Efficacy (absorption)	Agronomic biofortification (foliar)	NPNL women aged between 18 and 45 years	55	52–54%[a]
	Signorell et al. (2023)	India	Efficacy	Agronomic biofortification (foliar spray)	School aged children (4–12 years)	273	169%
Pearl millet	Mehta et al. (2022)	India	Efficacy	Iron- and zinc-biofortified pearl millet cultivar (*Dhanashakti*, ICTP-8203Fe)	Children aged 12–18 months	223	220%

Rice	Jongstra et al. (2022)	Bangladesh	Unclear	Conventionally bread high zinc strain (BRRI42) + agronomic (zinc foliar spraying)	Children aged 12–36-months	523	85%
Maize	Chomba et al. (2015)	Zambian	Efficacy (absorption)	Conventionally bread maize variety	Children aged 1–5 years	60	117%

[a]Depending on the extraction rate
NPNL non-pregnant non-lactating

zinc content of the food. The authors concluded that supplying biofortified maize can meet zinc requirements and provide an effective dietary substitute for regular maize for young children.

In a double-blind intervention trial, 1 to 3-year-old rural Bangladeshi children (n = 530), most of whom exhibited zinc-deficiency and stunted growth, were recruited and randomly assigned to receive either control rice (non-biofortified) or the biofortified variety for 9 months. While there was no significant difference between the amounts of rice consumed by the two groups (control: 232.7 ± 49.8 g/d; biofortified: 239.1 ± 43.4 g/d), the average daily zinc intake from the study rice was 1.20 ± 0.34 mg for the control group and 2.22 ± 0.47 mg for the biofortified group. However, this additional 1 mg per day of zinc did not translate into improvements in plasma zinc status, growth, or zinc-related morbidity among the participants (Jongstra et al. 2022).

Small-scale trials, including absorption studies primarily conducted on non-pregnant non-lactating (NPNL) women, suggest that the consumption of zinc-biofortified wheat results in an additional daily intake of approximately 2.5 mg to 6 mg, depending on flour extraction rates. Higher intake is observed with greater extraction rates (Lowe et al. 2021; Rosado et al. 2009; Signorell et al. 2019). Considering that roughly 75% of dietary zinc in predominantly wheat consuming populations is derived from wheat (as these studies calculated additional intake from bread rather than full meals), substituting the zinc biofortified variety for standard varieties can potentially fulfil 57–115% of the required daily intake (12.7 mg) for adult NPNL women (EFSA Panel on Dietetic Products, Nutrition and Allergies 2014).

It is noteworthy to observe that the study by Signorell et al. (2019) which explored the impact of the two agronomic approaches to biofortification on human zinc absorption found no discernible disparity in zinc absorption (both fractional and total absorbed zinc) in relation to food derived from wheat biofortified via foliar application or hydroponic root enrichment. Concurrently, absorption from the biofortified foods, irrespective of the agronomic biofortification technique employed, exhibited higher net zinc absorption in comparison to the control. Similarly, Rosado et al. (2009) also reported that net absorption from meals (2.1 ± 0.7 for 95% extraction and 2.0 ± 0.4 for 85% extraction) consisting of biofortified tortillas was 0.5 mg higher than from the non-biofortified control (1.6 ± 0.4 for 95% extraction and 1.5.0 ± 0.5 for 85% extraction). These values agreed well those predicted by an equation-based zinc absorption model that predicted 0.6 mg or 0.7 mg additional absorption from fortified meals made using 95% extraction or 80% extraction respectively, compared to the control wheat flour.

A recently published study (Signorell et al. 2023) presents findings from a 20-week double-blind intervention trial involving children aged 4 to 12 years (n = 273). The aim of the trial was to compare the effects of chapati made from agronomically biofortified whole wheat flour (BFW) on PZC when integrated into a mid-day school meal scheme. The study also included fortified control wheat (PHFW) and unfortified control wheat (CW) groups. The results revealed that the mean daily zinc intakes for the study groups BFW, PHFW, and CW groups were

4.4 ± 1.6, 5.9 ± 1.9, and 2.6 ± 0.6 mg of zinc per day, respectively. It is worth noting that these intakes were based on providing just one meal per day, and in the case of "universal biofortification," the zinc intake would likely be more substantial over time.

There was no significant difference in zinc intake between the PHFW and BFW groups, but both were significantly higher than the CW group. Despite the additional daily zinc intake of approximately 1.8 to 3.3 mg when consuming either PHFW or BFW as a single school meal per day, this did not lead to a positive effect on PZC, growth, or morbidity when compared to the control group. The study also included an additional plasma zinc analysis conducted four months after the intervention endpoint to understand the development of PZC post-intervention. In contrast to the PHFW and CW groups, which exhibited lower final PZC values compared to the measurements taken at the end of the intervention, the BFW group did not demonstrate a lower final PZC. This observation is intriguing and warrants further investigation.

Outcomes from two large-scale trials investigating the effectiveness of biofortified wheat have been published thus far. The first conducted in India was a community-based, double-masked randomized controlled trial (RCT) involving 6050 mother-child dyads (Sazawal et al. 2018). A standard commercial variety, PBW 550, was cultivated using standard farming techniques except that the high zinc wheat (HZn) for the intervention group was grown under specific agro-ecological conditions and received additional foliar spraying of 0.5% zinc sulfate fertilizer to enhance zinc uptake by the plants and its deposition, while 'Low zinc wheat' (LZn) control was grown in agro-ecological conditions that limit soil zinc uptake by plants and did not receive any additional zinc fertilization. Zinc content for the agronomically biofortified HZn and control LZn wheat flour was 30 mg/kg and 20 mg/kg, respectively. Participants received either HZn biofortified wheat flour or non-biofortified wheat flour for six months. Mothers enrolled in the study were NPNL women of reproductive age (15–49 years) and children were 4–6 years old. The study reported that compliance with consuming at least half of the recommended intake of flour was approximately 88% days for both women and children, while compliance with consuming the entire recommended amount (350 g for women and 120 g for children) was about 55% days. The zinc biofortified flour was estimated to deliver 3.6 mg/day of zinc to children compared with 2.4 mg/day for the control, providing a differential of 1.2 mgZn/day, and 10.8 mg/day to women compared to 7.8 mg/day for the control, providing a differential of 3 mgZn/day when the complete the recommended intakes was consumed by both the population subgroups.

The second trial was conducted in Pakistan (BiZiFED2) and had a cluster-randomized, double-blind, controlled design to understand the effectiveness of consuming zinc-biofortified wheat flour on the haematological indices of zinc in 517 adolescent girls (aged 10–16 years) in rural Pakistan under real-world scenarios (Lowe et al. 2020; Gupta et al. 2022a). In this study, the biofortified grain, grown by local farmers met the target zinc concentration of >40 mg/kg, averaging 45.3 mg/kg, with some variability (24.3 to 76.3 mg/kg). The provision of flour was made for the entire household, ensuring that family meals using the flour were consumed by the adolescent girls as part of their usual family meals. Similar to the study in India, the intervention lasted approximately 6 months, with the bread from biofortified flour providing 6.9 mgZn/day to the adolescent girls compared with 8.4 mgZn/day from the control flour, providing an average differential intake of 1.5 mgZn/day. The differential was less than expected due to the higher-than-average zinc content of the control flour. An earlier study by the same group, where both the control and biofortified wheat was grown under controlled conditions, reported an intake of 5.5 mgZn/day from biofortified and 2.5 mgZn/day from control flour, giving a differential intake of 3 mgZn/day for NPNL women for a comparable, low extraction (white) flour.

Neither study reported any significant difference in plasma zinc concentration between the intervention and control arms. The study conducted in Pakistan failed to show any intervention effect on linear growth and morbidity for adolescent girls and young children (secondary outcomes), although there was some indication of beneficial effects of the intervention on the incidence of respiratory tract infections towards the end of the study for both the population groups (Gupta et al. 2022b; Gupta et al. 2023). Additionally, this study in Pakistan also found a modest increase in head circumference among children in biofortified group compared to control (unpublished). In India, biofortification showed positive impacts on self-reported morbidity among both the population groups (Sazawal et al. 2018).

Beside the two studies described above, several recent ex-ante studies have examined how biofortified crops impact human well-being using the disability-adjusted life year (DALY) method. The first study focused on the potential health benefits of golden rice in the Philippines (Zimmermann and Qaim 2004). This approach was later expanded to include crops fortified with zinc and iron, and was applied across various countries (Stein et al. 2005). A more recent analysis (Liu et al. 2017) evaluated the effect of agronomic biofortification (via application of six rates of zinc fertilizer to soil) on zinc bioavailability in wheat grain and flour and its impacts on human health using DAILY approach. Zinc bioavailability was estimated using a mathematical model. It showed that the zinc concentration increased in all flour fractions with an increase in rate of zinc fertilization, however the percentages of zinc in standard flour (25%) and bran (75%) relative to total grain zinc were constant. Phytic acid concentrations in grain and flours were unaffected by zinc biofortification. The availability of zinc and its impact on health as measured by saved DALYs, escalated with the zinc application rate. This effect was more pronounced in standard white flour, and highly processed refined flour compared to whole grain and coarse flour. Standard and refined flour from biofortified sources, achieved through agronomic methods, met the target of 3 mg of zinc from 300 g of wheat flour and led to a >20% reduction in DALYs.

Overall, the above studies including ex-ante evaluations and feeding studies provide evidence that traditional breeding and agronomic methods of biofortification led to significantly increased dietary zinc intake compared with controls, without compromising bioavailability.

Forging Ahead with Zinc Biofortification: Navigating Challenges and Impact on Human Health

In general, biofortification has three core requisites for its success (1) Effective breeding that involves combining high nutrient density with substantial yields and profitability; (2) Biofortified crops must gain traction among farmers, with their grain reaching those most vulnerable to micronutrient malnutrition; (3) Ability to demonstrate efficacy by showcasing improved micronutrient status and/or related health outcomes through consuming biofortified varieties within the usual diet (Listman et al. 2019). In the case of zinc biofortification using conventional and agronomic techniques, it is evident that breeding has been effective and can provide several other benefits, such as high yield, improved seed and seedling vigour, reduced root and shoot accumulation of cadmium, as well as offering resistance towards certain pest and pathogens. Thus, the "invisible" nutrient zinc, when integrated into resilient high yielding varieties, acts synergistically to provide 'added market value' for farmers to incentivise adoption. Further to this, concerted efforts from the governments and non-government agencies have facilitated the release and scale up of zinc rich cereals in various regions of the world, in particular South-Asia, where the greatest impact of zinc deficiencies including impaired childhood growth, morbidity and mortality, and adverse maternal health and pregnancy outcomes are witnessed.

Although assessments of intakes through limited human studies and ex-ante evaluations suggest that zinc biofortification of cereals can enhance zinc intake, empirical data supporting its translation into human health benefits remain fragmented, making it challenging to draw definitive conclusions. It is crucial to be able to directly measure changes in the prevalence of deficiency resulting from the consumption of biofortified staples, which necessitates controlled trials to validate the impact at achieved nutrient density levels. Zinc deficiency is associated with impaired growth and immunity (King et al. 2015; Black and Sazawal 2001; Liu et al. 2018). In fact, the percentage of children <5 years of age with height-for-age Z scores (HAZ) below -2 SD of the WHO reference has also been suggested as a proxy indicator for assessing at-risk populations and initiating program planning for zinc interventions (de Benoist et al. 2007). Consequently, several cereal biofortification trials also include height and derived HAZ, as well as zinc-related morbidities such as diarrhoea and respiratory tract infections, as outcome variables alongside PZC.

The effectiveness and efficacy studies done so far have often yielded contradictory results concerning zinc status. The sole study conducted on rice in Bangladesh failed to demonstrate any effect on PZC of consuming zinc-biofortified rice for 9 months which provided approximately 1 mg of additional zinc daily, or on the prevalence of zinc deficiency based on PZC (Jongstra et al. 2022). This intervention had no significant effect on diarrhoea. Surprisingly, an 8% higher overall morbidity rate was reported in the intervention group due to a higher incidence of upper respiratory tract illnesses in this group. Since no reasonable explanation could be identified for this observation, the authors attributed this outcome to coincidence.

The study by Mehta and co-workers (2022) concluded the daily consumption of iron-zinc pearl millet -based complementary foods did not significantly impact iron and zinc status or growth in children living in an urban slum of western India. However, it primarily evaluated the effect on iron status and therefore the sub-group analysis conducted only for iron status, indicated improved haemoglobin concentrations among male children and among individuals who were iron-deficient or iron-depleted at baseline.

All the studies carried out to understand the usefulness of biofortified wheat consumption for improving zinc status consistently fail to exhibit any positive impact on PZC (Gupta et al. 2022a; Sazawal et al. 2018; Signorell et al. 2023) except a short duration 8-week intervention in Pakistan (Lowe et al. 2021). In this study, although a significant increase in plasma zinc concentration after 4 weeks was observed in the intervention arm but not control, this effect was not sustained at 8 weeks which marked the intervention endpoint.

While it could be contended that the additional absorbed zinc from biofortified flour might have limited impact on PZC due to high phytate intake among the participants or modest zinc increments from biofortified meal, it important to acknowledge the well-known constraints of PZC as an indicator (King et al. 2015). PZC is a common measure for evaluating zinc status in populations, however, it is homeostatically controlled and at an individual level, thus responses to modest dietary zinc changes, are subtle (King 2011). This is especially true when the extra zinc is ingested with food, such as from consuming biofortified staples rather as a supplement. Also, challenges in interpreting PZC arise from factors such as concurrent infection, fasting, non-fasting states, and the time of day (McDonald et al. 2020; Arsenault et al. 2011). Considering these limitations of PZC, several novel biomarkers, including products of essential fatty acid metabolism, DNA fragmentation, hair and nail zinc content, are being tested in real-world settings and may enable the effect of modest changes in dietary zinc intake via biofortification to be monitored (Lowe et al. 2020; Signorell et al. 2023; Jongstra et al. 2022; Zyba et al. 2017; Liong et al. 2021). A study nested in the BiZiFED2 effectiveness trial reported a detectable increase zinc counts, adjusted for sulfur (Zn:S count ratio) in individual hair strands, measured using X-ray fluorescence spectrometry, in response to a modest increase in dietary zinc (1.5 mg/day) over 6 months among adolescent Pakistani girls aged 10–16 years (Frederickson et al. 2023). Such methods offer a sensitive, non-invasive method to monitor changes within subjects in response to dietary zinc interventions and should be further tested for robustness in free-living, community settings where confounding co-morbidities may be present.

In terms of functional indicators, studies have generally failed to demonstrate any measurable impact of consuming biofortified foods on anthropometric data. However, there are some findings indicating a positive impact of consuming zinc-biofortified wheat on self-reported morbidities. In 2018, research conducted in India reported that young children who included zinc-biofortified wheat in their diets, commonly consumed through items such as chapatis, puri (flatbreads), or porridge, exhibited a notable 17% decrease in the frequency of days they experienced pneumonia, along with a substantial 39% reduction in the number of days they encountered vomiting, when compared to children who consumed conventional wheat-based products over a period of 6 months (Sazawal et al. 2018).

In the BiZiFED 2 study, a lower incidence of respiratory tract infections (RTIs) was reported in the intervention arm compared to the control arm at the end of the 25-week intervention period, during which the biofortified group of adolescent girls consumed an additional 1.5 mg of dietary zinc daily compared to the control group (Gupta et al. 2022b). Similar intervention effects on incidences of RTI were reported for young children (1-5 years) in this study (Gupta et al. 2023). However, when considering the longitudinal prevalence of RTIs (cumulative days of sickness as a percentage of total observation days) with baseline adjustments, no differences between the groups were observed in either population. The duration of large-scale intervention studies are inevitably limited by cost and resource. The ongoing scale-up of the release of zinc-biofortified cereals provides an opportunity to conduct longer-term (>1 year) observational studies to monitor changes in such functional outcomes over time.

Interaction between phytate and zinc presents another critical impediment in realizing the full potential of the biofortification strategy. Phytic acid, a naturally occurring compound in many plant-based foods, has the ability to form insoluble complexes (phytate) with minerals such as zinc, rendering them less available for absorption by the human body (Lönnerdal 2000; Gibson et al. 2010). Scientists are researching strategies such as low-phytate crops and processing techniques to mitigate this. Evidence suggests that plant breeding techniques hold promise for enhancing zinc bioavailability as well. Previously, maize with low phytate content, developed through plant breeding, showed improved zinc absorption in short-term studies (Adams et al. 2002; Hambidge et al. 2004). However, a longer-term study providing low-phytate maize to Guatemalan schoolchildren was unsuccessful in establishing enhanced zinc absorption compared to control maize (Mazariegos et al. 2006). The reasons behind these unexpected results remain unclear. Validating the long-term effectiveness of low-phytate hybrids is essential, as this approach could significantly improve absorbable zinc intake for populations relying on plant-based diets. Challenges such as reduced yields associated with the low-phytate trait and the need for dedicated long-term breeding projects have hindered further exploration of this strategy.

Studies in wheat and other cereals have shown that transgenic strategies can be used to increase the contents of iron and zinc in white flour by converting the starchy endosperm tissue into a 'sink' for minerals (Harrington et al. 2022). Although such strategies currently have low acceptability, a greater understanding of the mechanisms that control the transport and deposition of iron and zinc in the developing grain should allow similar effects to be achieved by exploiting naturally occurring genetic variation (Balk et al. 2019).

Mechanical treatments and fermentation are two of the most promising processing techniques. Many microorganisms secrete phytase enzymes, which can release minerals from phytate complexes, particularly microorganisms present in sourdough systems (Lopez et al. 2003; Rodriguez-Ramiro et al. 2017). Hence, sourdough wholegrain products may have increased mineral bioavailability. However, this approach may increase mineral bioavailability in foods made from wholegrain and high-extraction flours, but it is not relevant to white flour products, which are preferred in most countries. Micro-milling, a processing technique whereby aleurone cell walls (containing 70% of grain iron and zinc) are ruptured, can increase the availability of minerals from wheat flour. A study to explore whether micro-milling can increase iron and zinc availability from biofortified wholegrain flour as well as from aleurone-enriched white flour is underway (UK Research and Innovation (UKRI) 2016). If successful, such strategies are expected to enhance the mineral absorption potential across various wheat-based products, ranging from refined flour to whole-grain products.

Conclusion

Overall, within the context of cereals, particularly maize and wheat discussed here, the concept of zinc biofortification emerges as a promising strategy for improving nutritional status on a population scale. Zinc biofortification enhances zinc content, yield, and resistance to various pests, which encourages adoption. It undeniably demonstrates its success in increasing zinc intakes in various population sub-groups. However, despite increased zinc intake, its translation into better health is inconclusive, primarily due to a lack of sensitive and reliable biomarkers. Novel biomarkers, such as single hair analysis by X-ray fluorescence spectrometry may offer greater sensitivity and need to be tested alongside widely used PZC. Long-term interventions are warranted to further confirm positive findings related to self-reported morbidities and assess the impact on growth among children.

References

Adams, C. L., Hambidge, M., Raboy, V., Dorsch, J. A., Sian, L., Westcott, J. L., et al. (2002). Zinc absorption from a low-phytic acid maize. *The American Journal of Clinical Nutrition, 76*(3), 556–559. https://doi.org/10.1093/ajcn/76.3.556

Ansari, N., Mehmood, R., & Gazdar, H. (2018). Going against the grain of optimism: Flour fortification in Pakistan. *IDS Bulletin, 49*, 57–71. https://doi.org/10.19088/1968-2018.104

Arsenault, J. E., Wuehler, S. E., de Romaña, D. L., Penny, M. E., Sempértegui, F., & Brown, K. H. (2011). The time of day and the interval since previous meal are associated with plasma zinc concentrations and affect estimated risk of zinc deficiency in young children in Peru and Ecuador. *European Journal of Clinical Nutrition, 65*(2), 184–190. https://doi.org/10.1038/ejcn.2010.234

Bailey, R. L., West, K. P., Jr., & Black, R. E. (2015). The epidemiology of global micronutrient deficiencies. *Annals of Nutrition and Metabolism, 66*(Suppl. 2), 22–33. https://doi.org/10.1159/000371618

Balk, J., Connorton, J. M., Wan, Y., Lovegrove, A., Moore, K. L., Uauy, C., et al. (2019). Improving wheat as a source of iron and zinc for global nutrition. *Nutrition Bulletin, 44*(1), 53–59. https://doi.org/10.1111/nbu.12361

Bhardwaj, A. K., Chejara, S., Malik, K., Kumar, R., Kumar, A., & Yadav, R. K. (2022). Agronomic biofortification of food crops: An emerging opportunity for global food and nutritional security. *Frontiers in Plant Science, 13*, 1055278. https://doi.org/10.3389/fpls.2022.1055278

Bhutta, Z., Soofi, S., Zaidi, S., Habib, A., Hussain, Bhutta, Z., Soofi, S., Zaidi, S., Habib, A., & Hussain, I. (2011). *Pakistan National Nutrition Survey, 2011*. Available at: https://ecommons.aku.edu/pakistan_fhs_mc_women_childhealth_paediatr/262. Accessed on 21.09.2023.

Black, R. E. (2019). Progress in the use of ORS and zinc for the treatment of childhood diarrhea. *Journal of Global Health, 9*(1), 010101. https://doi.org/10.7189/jogh.09.010101

Black, R. E., & Sazawal, S. (2001). Zinc and childhood infectious disease morbidity and mortality. *The British Journal of Nutrition, 85*(Suppl 2), S125–S129. https://doi.org/10.1079/bjn2000304

Bouis, H. E., & Saltzman, A. (2017). Improving nutrition through biofortification: A review of evidence from HarvestPlus, 2003 through 2016. *Global Food Security, 12*, 49–58. https://doi.org/10.1016/j.gfs.2017.01.009

Brown, K. H., Wuehler, S. E., & Peerson, J. M. (2001). The importance of zinc in human nutrition and estimation of the global prevalence of zinc deficiency. *Food and Nutrition Bulletin, 22*(2), 113–125. https://doi.org/10.1177/156482650102200201

Brown, K. H., Rivera, J. A., Bhutta, Z., Gibson, R. S., King, J. C., Lönnerdal, B., et al. (2004). International Zinc Nutrition Consultative Group (IZiNCG) technical document #1. Assessment of the risk of zinc deficiency in populations and options for its control. *Food and Nutrition Bulletin, 25*(1 Suppl 2), S99–S203.

Cakmak, I., & Kutman, U. B. (2018). Agronomic biofortification of cereals with zinc: A review. *European Journal of Soil Science, 69*(1), 172–180. https://doi.org/10.1111/ejss.12437

Caulfield, L. E., & Black, R. E. (2004). Zinc deficiency. In *Comparative quantification of health risks: Global and regional burden of disease attributable to selected major risk factors* (Vol. 1, pp. 257–280).

Ceballos-Rasgado, M., Moran, V., Ander, E. L., Ajmal, S., Joy, E., Mahboob, U., et al. (2022). Acceptability of zinc biofortified wheat and flour among farmers in Pakistan: Experiences from the BiZiFED2 project. *Proceedings of the Nutrition Society, 81*(OCE5), E178. https://doi.org/10.1017/S0029665122002117

Chomba, E., Westcott, C. M., Westcott, J. E., Mpabalwani, E. M., Krebs, N. F., Patinkin, Z. W., et al. (2015). Zinc absorption from biofortified maize meets the requirements of young rural Zambian children. *The Journal of Nutrition, 145*(3), 514–519. https://doi.org/10.3945/jn.114.204933

Cui, K., & Shoemaker, S. P. (2018). Public perception of genetically-modified (GM) food: A nationwide Chinese Consumer Study. *npj Science of Food, 2*(1), 10. https://doi.org/10.1038/s41538-018-0018-4

Dapkekar, A., Deshpande, P., Oak, M. D., Paknikar, K. M., & Rajwade, J. M. (2018). Zinc use efficiency is enhanced in wheat through nanofertilization. *Scientific Reports, 8*(1), 6832. https://doi.org/10.1038/s41598-018-25247-5

de Benoist, B., Darnton-Hill, I., Davidsson, L., Fontaine, O., & Hotz, C. (2007). Conclusions of the joint WHO/UNICEF/IAEA/IZiNCG interagency meeting on zinc status indicators. *Food and Nutrition Bulletin, 28*(3 Suppl), S480–S484. https://doi.org/10.1177/15648265070283s306

Dubock, A. (2017). An overview of agriculture, nutrition and fortification, supplementation and biofortification: Golden Rice as an example for enhancing micronutrient intake. *Agriculture & Food Security, 6*(1), 59. https://doi.org/10.1186/s40066-017-0135-3

EFSA Panel on Dietetic Products, Nutrition and Allergies. (2014). Scientific opinion on dietary reference values for zinc. *EFSA Journal, 12*(10), 3844.

e-Pact Consortium. (2019). *Evaluation of the supporting nutrition in Pakistan Food Fortification Programme. Midterm evaluation report*. Available online: https://www.opml.co.uk/files/2021-07/mid-term-evaluation-report-food-fortification-programme.pdf?noredirect=1. Accessed on 21.09.2023.

e-Pact Consortium. (2021). *Evaluation of the supporting nutrition in Pakistan Food Fortification Programme endline evaluation report.* Available at: https://iati.fcdo.gov.uk/iati_documents/90000152.pdf. Accessed on 21.09.2023.

Fischer Walker, C. L., Ezzati, M., & Black, R. E. (2009). Global and regional child mortality and burden of disease attributable to zinc deficiency. *European Journal of Clinical Nutrition, 63*(5), 591–597. https://doi.org/10.1038/ejcn.2008.9

Food and Agriculture Organization of United Nation. (2023). *Food and agriculture database.* https://www.fao.org/faostat/en/#home. Accessed on 27.09.2023.

Frederickson, C. J., Fleming, D. E. B., Asael, D., Zaman, M., Ferguson, R., Kaiser, M. G., et al. (2023). Single hair analysis by X-ray fluorescence spectrometry detects small changes in dietary zinc intake: A nested randomized controlled trial. *Frontiers in Nutrition, 10,* 1139017. https://doi.org/10.3389/fnut.2023.1139017

Gaikwad, K. B., Rani, S., Kumar, M., Gupta, V., Babu, P. H., Bainsla, N. K., et al. (2020). Enhancing the nutritional quality of major food crops through conventional and genomics-assisted breeding. *Frontiers in Nutrition, 7,* 533453. https://doi.org/10.3389/fnut.2020.533453

Gannon, B. M., Thakker, V., Bonam, V. S., Haas, J. D., Bonam, W., Finkelstein, J. L., et al. (2019). A randomized crossover study to evaluate recipe acceptability in breastfeeding mothers and young children in India targeted for a multiple biofortified food crop intervention. *Food and Nutrition Bulletin, 40*(4), 460–470. https://doi.org/10.1177/0379572119855588

Garg, M., Sharma, N., Sharma, S., Kapoor, P., Kumar, A., Chunduri, V., et al. (2018). Biofortified crops generated by breeding, agronomy, and transgenic approaches are improving lives of millions of people around the world. *Frontiers in Nutrition, 5,* 12. https://doi.org/10.3389/fnut.2018.00012

Gibson, R. S., & Anderson, V. P. (2009). A review of interventions based on dietary diversification or modification strategies with the potential to enhance intakes of total and absorbable zinc. *Food and Nutrition Bulletin, 30*(1 Suppl), S108–S143. https://doi.org/10.1177/15648265090301s107

Gibson, R. S., & Ferguson, E. L. (1998). Nutrition intervention strategies to combat zinc deficiency in developing countries. *Nutrition Research Reviews, 11*(1), 115–131. https://doi.org/10.1079/NRR19980008

Gibson, R. S., Bailey, K. B., Gibbs, M., & Ferguson, E. L. (2010). A review of phytate, iron, zinc, and calcium concentrations in plant-based complementary foods used in low-income countries and implications for bioavailability. *Food and Nutrition Bulletin, 31*(2 Suppl), S134–S146. https://doi.org/10.1177/15648265100312s206

Government of Pakistan and United Nations Children's Fund (UNICEF). (2023). *Pakistan National Nutrition Survey 2018.* Available at: https://www.unicef.org/pakistan/reports/national-nutrition-survey-2018-full-report-3-volumes-key-findings-report. Accessed 21.09.2023.

Gupta, S., Brazier, A. K. M., & Lowe, N. M. (2020). Zinc deficiency in low- and middle-income countries: Prevalence and approaches for mitigation. *Journal of Human Nutrition and Dietetics, 33*(5), 624–643. https://doi.org/10.1111/jhn.12791

Gupta, S., Zaman, M., Fatima, S., Shahzad, B., Brazier, A. K. M., Moran, V. H., et al. (2022a). The impact of consuming zinc-biofortified wheat flour on haematological indices of zinc and iron status in adolescent girls in rural Pakistan: A cluster-randomised, double-blind, controlled effectiveness trial. *Nutrients, 14*(8). https://doi.org/10.3390/nu14081657

Gupta, S., Shahzad, B., Zaman, M., Sinclair, J., Fatima, S., Brazier, A., et al. (2022b). Impact of consuming zinc-biofortified wheat flour on the growth and morbidity status of adolescent girls: A cluster randomised, double blind, controlled trial. *Proceedings of the Nutrition Society, 81*(OCE5), E171. https://doi.org/10.1017/S002966512200204X

Gupta, S., Zaman, M., Fatima, S., Moran, V. H., Sinclair, J. K., & Nicola, M. (2023). Lowe Impact of consuming zinc-biofortified wheat flour on the growth and morbidity status of children aged 1–5 years: A cluster-randomized, double-blind, controlled trial. In *Proceedings of the Nutrition Society.* Accepted for publication.

Hambidge, K. M., Huffer, J. W., Raboy, V., Grunwald, G. K., Westcott, J. L., Sian, L., et al. (2004). Zinc absorption from low-phytate hybrids of maize and their wild-type isohybrids. *The American Journal of Clinical Nutrition, 79*(6), 1053–1059. https://doi.org/10.1093/ajcn/79.6.1053

Harrington, S. A., Connorton, J. M., Nyangoma, N. I. M., McNelly, R., Morgan, Y. M. L., Aslam, M. F., et al. (2022). A two-gene strategy increases iron and zinc concentrations in wheat flour, improving mineral bioaccessibility. *Plant Physiology, 191*(1), 528–541. https://doi.org/10.1093/plphys/kiac499

HarvestPlus. (2022). *The journey of scaling in Pakistan: Zinc-enriched biofortified wheat*. Available at: https://www.harvestplus.org/wp-content/uploads/2022/03/The-Journey-of-Scaling-in-Pakistan.pdf. Accessed on 21.09.2023.

HarvestPlus. (2023). *HarvestPlus Database of Biofortified crop released*. Available at: https://bcr.harvestplus.org/. Acessed on 27.09.2023.

Jongstra, R., Hossain, M. M., Galetti, V., Hall, A. G., Holt, R. R., Cercamondi, C. I., et al. (2022). The effect of zinc-biofortified rice on zinc status of Bangladeshi preschool children: A randomized, double-masked, household-based, controlled trial. *The American Journal of Clinical Nutrition, 115*(3), 724–737. https://doi.org/10.1093/ajcn/nqab379

King, J. C. (2011). Zinc: an essential but elusive nutrient. *The American Journal of Clinical Nutrition, 94*(2), 679s–684s. https://doi.org/10.3945/ajcn.110.005744

King, J. C., Brown, K. H., Gibson, R. S., Krebs, N. F., Lowe, N. M., Siekmann, J. H., et al. (2015). Biomarkers of nutrition for development (BOND)-zinc review. *The Journal of Nutrition, 146*(4), 858s–885s. https://doi.org/10.3945/jn.115.220079

Kumar, A., Mehtre, S., Anuradha, K., Jaganathan, J., Prasanna, H., Gorthy, S., et al. (2018). Delivering bioavailable micronutrients through biofortifying sorghum and seed chain innovations. *Science Forums*. Available at: https://oar.icrisat.org/10924/

Kumar, K., Gambhir, G., Dass, A., Tripathi, A. K., Singh, A., Jha, A. K., et al. (2020). Genetically modified crops: Current status and future prospects. *Planta, 251*(4), 91. https://doi.org/10.1007/s00425-020-03372-8

Liong, E. M., McDonald, C. M., Suh, J., Westcott, J. L., Wong, C. P., Signorell, C., et al. (2021). Zinc-biofortified wheat intake and zinc status biomarkers in men: Randomized controlled trial. *The Journal of Nutrition, 151*(7), 1817–1823. https://doi.org/10.1093/jn/nxab092

Listman, G. M., Guzmán, C., Palacios-Rojas, N., Pfeiffer, W. H., San Vicente, F., & Govindan, V. (2019). Improving nutrition through biofortification: Preharvest and postharvest technologies. *Cereal Foods World, 64*(3), 1–7.

Liu, D., Liu, Y., Zhang, W., Chen, X., & Zou, C. (2017). Agronomic approach of zinc biofortification can increase zinc bioavailability in wheat flour and thereby reduce zinc deficiency in humans. *Nutrients, 9*(5), 465.

Liu, E., Pimpin, L., Shulkin, M., Kranz, S., Duggan, C. P., Mozaffarian, D., et al. (2018). Effect of zinc supplementation on growth outcomes in children under 5 years of age. *Nutrients, 10*(3). https://doi.org/10.3390/nu10030377

Lockyer, S., White, A., & Buttriss, J. (2018). Biofortified crops for tackling micronutrient deficiencies – what impact are these having in developing countries and could they be of relevance within Europe? *Nutrition Bulletin, 43*, 319–357. https://doi.org/10.1111/nbu.12347

Lönnerdal, B. (2000). Dietary factors influencing zinc absorption. *The Journal of Nutrition, 130*(5), 1378S–1383S. https://doi.org/10.1093/jn/130.5.1378S

Lopez, H. W., Duclos, V., Coudray, C., Krespine, V., Feillet-Coudray, C., Messager, A., et al. (2003). Making bread with sourdough improves mineral bioavailability from reconstituted whole wheat flour in rats. *Nutrition, 19*(6), 524–530. https://doi.org/10.1016/s0899-9007(02)01079-1

Lowe, N. M. (2021). The global challenge of hidden hunger: Perspectives from the field. *Proceedings of the Nutrition Society, 80*(3), 283–289. https://doi.org/10.1017/S0029665121000902

Lowe, N. M., Khan, M. J., Broadley, M. R., Zia, M. H., McArdle, H. J., Joy, E. J. M., et al. (2018). Examining the effectiveness of consuming flour made from agronomically biofortified wheat (Zincol-2016/NR-421) for improving Zn status in women in a low-resource setting in Pakistan:

Study protocol for a randomised, double-blind, controlled cross-over trial (BiZiFED). *BMJ Open, 8*(4), e021364. https://doi.org/10.1136/bmjopen-2017-021364

Lowe, N. M., Zaman, M., Moran, V. H., Ohly, H., Sinclair, J., Fatima, S., et al. (2020). Biofortification of wheat with zinc for eliminating deficiency in Pakistan: Study protocol for a cluster-randomised, double-blind, controlled effectiveness study (BIZIFED2). *BMJ Open, 10*(11), e039231. https://doi.org/10.1136/bmjopen-2020-039231

Lowe, N. M., Zaman, M., Khan, M. J., Brazier, A. K. M., Shahzad, B., Ullah, U., et al. (2021). Biofortified wheat increases dietary zinc intake: A randomised controlled efficacy study of Zincol-2016 in rural Pakistan. *Frontiers in Nutrition, 8*, 809783. https://doi.org/10.3389/fnut.2021.809783

Mahboob, U., Ohly, H., Joy, E. J. M., Moran, V., Zaman, M., & Lowe, N. M. (2020). Exploring community perceptions in preparation for a randomised controlled trial of biofortified flour in Pakistan. *Pilot and Feasibility Studies, 6*(1), 117. https://doi.org/10.1186/s40814-020-00664-4

Mahboob, U., Ceballos-Rasgado, M., Moran, V. H., Joy, E. J. M., Ohly, H., Zaman, M., et al. (2022). Community perceptions of zinc biofortified flour during an intervention study in Pakistan. *Nutrients, 14*(4), 817. https://doi.org/10.3390/nu14040817

Maqbool, M. A., & Beshir, A. (2019). Zinc biofortification of maize (Zea mays L.): Status and challenges. *Plant Breeding, 138*(1), 1–28. https://doi.org/10.1111/pbr.12658

Mazariegos, M., Hambidge, K. M., Krebs, N. F., Westcott, J. E., Lei, S., Grunwald, G. K., et al. (2006). Zinc absorption in Guatemalan schoolchildren fed normal or low-phytate maize. *The American Journal of Clinical Nutrition, 83*(1), 59–64. https://doi.org/10.1093/ajcn/83.1.59

McDonald, C. M., Suchdev, P. S., Krebs, N. F., Hess, S. Y., Wessells, K. R., Ismaily, S., et al. (2020). Adjusting plasma or serum zinc concentrations for inflammation: Biomarkers Reflecting Inflammation and Nutritional Determinants of Anemia (BRINDA) project. *The American Journal of Clinical Nutrition, 111*(4), 927–937. https://doi.org/10.1093/ajcn/nqz304

Mehta, S., Huey, S. L., Ghugre, P. S., Potdar, R. D., Venkatramanan, S., Krisher, J. T., et al. (2022). A randomized trial of iron- and zinc-biofortified pearl millet-based complementary feeding in children aged 12 to 18 months living in urban slums. *Clinical Nutrition, 41*(4), 937–947. https://doi.org/10.1016/j.clnu.2022.02.014

Nasrin, D., Wu, Y., Blackwelder, W. C., Farag, T. H., Saha, D., Sow, S. O., et al. (2013). Health care seeking for childhood diarrhea in developing countries: Evidence from seven sites in Africa and Asia. *The American Journal of Tropical Medicine and Hygiene, 89*(1 Suppl), 3–12. https://doi.org/10.4269/ajtmh.12-0749

Nestel, P., Bouis, H. E., Meenakshi, J. V., & Pfeiffer, W. (2006). Biofortification of staple food crops. *The Journal of Nutrition, 136*(4), 1064–1067. https://doi.org/10.1093/jn/136.4.1064

Ohly, H., Broadley, M. R., Joy, E. J. M., Khan, M. J., McArdle, H., Zaman, M., et al. (2019). The BiZiFED project: Biofortified zinc flour to eliminate deficiency in Pakistan. *Nutrition Bulletin, 44*(1), 60–64. https://doi.org/10.1111/nbu.12362

Praharaj, S., Skalicky, M., Maitra, S., Bhadra, P., Shankar, T., Brestic, M., et al. (2021). Zinc biofortification in food crops could alleviate the zinc malnutrition in human health. *Molecules, 26*(12). https://doi.org/10.3390/molecules26123509

Prasad, A. S. (2013). Discovery of human zinc deficiency: Its impact on human health and disease. *Advances in Nutrition, 4*(2), 176–190. https://doi.org/10.3945/an.112.003210

Prasanna, B. M., Palacios-Rojas, N., Hossain, F., Muthusamy, V., Menkir, A., Dhliwayo, T., et al. (2019). Molecular breeding for nutritionally enriched maize: Status and prospects. *Frontiers in Genetics, 10*, 1392. https://doi.org/10.3389/fgene.2019.01392

Rashid, A., Ram, H., Zou, C.-Q., Rerkasem, B., Duarte, A. P., Simunji, S., et al. (2019). Effect of zinc-biofortified seeds on grain yield of wheat, rice, and common bean grown in six countries. *Journal of Plant Nutrition and Soil Science, 182*(5), 791–804. https://doi.org/10.1002/jpln.201800577

Rizwan, M., Zhu, Y., Qing, P., Zhang, D., Ahmed, U. I., Xu, H., et al. (2021). Factors determining consumer acceptance of biofortified food: Case of zinc-fortified wheat in Pakistan's Punjab Province. *Frontiers in Nutrition, 8*, 647823. https://doi.org/10.3389/fnut.2021.647823

Rodriguez-Ramiro, I., Brearley, C. A., Bruggraber, S. F., Perfecto, A., Shewry, P., & Fairweather-Tait, S. (2017). Assessment of iron bioavailability from different bread making processes using an in vitro intestinal cell model. *Food Chemistry, 228*, 91–98. https://doi.org/10.1016/j.foodchem.2017.01.130

Rosado, J. L., Hambidge, K. M., Miller, L. V., Garcia, O. P., Westcott, J., Gonzalez, K., et al. (2009). The quantity of zinc absorbed from wheat in adult women is enhanced by biofortification. *The Journal of Nutrition, 139*(10), 1920–1925. https://doi.org/10.3945/jn.109.107755

Saha, I., Durand-Morat, A., Nalley, L. L., Alam, M. J., & Nayga, R. (2021). Rice quality and its impacts on food security and sustainability in Bangladesh. *PLoS One, 16*(12), e0261118. https://doi.org/10.1371/journal.pone.0261118

Saltzman, A., Birol, E., Bouis, H. E., Boy, E., De Moura, F. F., Islam, Y., et al. (2013). Biofortification: Progress toward a more nourishing future. *Global Food Security, 2*(1), 9–17. https://doi.org/10.1016/j.gfs.2012.12.003

Sandstead, H. H. (2000). Causes of iron and zinc deficiencies and their effects on brain. *The Journal of Nutrition, 130*(2S Suppl), 347s–349s. https://doi.org/10.1093/jn/130.2.347S

Sazawal, S., Dhingra, U., Dhingra, P., Dutta, A., Deb, S., Kumar, J., et al. (2018). Efficacy of high zinc biofortified wheat in improvement of micronutrient status, and prevention of morbidity among preschool children and women – a double masked, randomized, controlled trial. *Nutrition Journal, 17*(1), 86. https://doi.org/10.1186/s12937-018-0391-5

Signorell, C., Zimmermann, M. B., Cakmak, I., Wegmüller, R., Zeder, C., Hurrell, R., et al. (2019). Zinc absorption from agronomically biofortified wheat is similar to post-harvest fortified wheat and is a substantial source of bioavailable zinc in humans. *The Journal of Nutrition, 149*(5), 840–846. https://doi.org/10.1093/jn/nxy328

Signorell, C., Kurpad, A. V., Pauline, M., Shenvi, S., Mukhopadhyay, A., King, J. C., et al. (2023). The effect of zinc biofortified wheat produced via foliar application on zinc status: A randomized, controlled trial in Indian children. *The Journal of Nutrition, 153*, 3092–3100. https://doi.org/10.1016/j.tjnut.2023.08.013

Sitaresmi, T., Hairmansis, A., Widyastuti, Y., Rachmawati, S. U., Wibowo, B. P., et al. (2023). Advances in the development of rice varieties with better nutritional quality in Indonesia. *Journal of Agriculture and Food Research, 12*, 100602. https://doi.org/10.1016/j.jafr.2023.100602

Stein, A. J., Meenakshi, J. V., Qaim, M., Nestel, P., & Bhutta, Z. (2005). Analyzing the health benefits of biofortified staple crops by means of the disability-adjusted life years approach. In *A handbook focusing on iron, zinc and vitamin A* (Harvest Plus Technical Monograph Series) (pp. 1–32). Food and Agriculture Organization of the United Nations.

Stevens, G. A., Beal, T., Mbuya, M. N. N., Luo, H., & Neufeld, L. M. (2022). Micronutrient deficiencies among preschool-aged children and women of reproductive age worldwide: A pooled analysis of individual-level data from population-representative surveys. *The Lancet Global Health, 10*(11), e1590–e15e9. https://doi.org/10.1016/s2214-109x(22)00367-9

Talsma, E. F., Melse-Boonstra, A., & Brouwer, I. D. (2017). Acceptance and adoption of biofortified crops in low- and middle-income countries: A systematic review. *Nutrition Reviews, 75*(10), 798–829. https://doi.org/10.1093/nutrit/nux037

Tulchinsky, T. H. (2010). Micronutrient deficiency conditions: Global health issues. *Public Health Reviews, 32*(1), 243–255. https://doi.org/10.1007/BF03391600

UK Research and Innovation (UKRI). (2016). *Increasing micronutrient bioavailability from wheat*. https://gtr.ukri.org/projects?ref=BB/N021002/1. Accessed 27.09.2023.

Wani, S. H., Gaikwad, K., Razzaq, A., Samantara, K., Kumar, M., & Govindan, V. (2022). Improving zinc and iron biofortification in wheat through genomics approaches. *Molecular Biology Reports, 49*(8), 8007–8023. https://doi.org/10.1007/s11033-022-07326-z

Wiafe, M. A., Apprey, C., & Annan, R. A. (2023). Dietary diversity and nutritional status of adolescents in rural Ghana. *Nutrition and Metabolic Insights, 16*, 11786388231158487. https://doi.org/10.1177/11786388231158487

Woods, B. J., Gallego-Castillo, S., Talsma, E. F., & Álvarez, D. (2020). The acceptance of zinc biofortified rice in Latin America: A consumer sensory study and grain quality characterization. *PLoS One, 15*(11), e0242202. https://doi.org/10.1371/journal.pone.0242202

World Health Organization (WHO), Johns Hopkins Bloomberg School of Public Health, United States Agency for International Development (USAIDS), United Nations Children's Fund (UNICEF). (2006). *Implementing the new recommendations on the clinical management of diarrhoea: Guidelines for policy makers and programme managers*. World Health Organization. Available at: https://iris.who.int/handle/10665/43456. Accessed online 27.09.2023.

Zimmermann, R., & Qaim, M. (2004). Potential health benefits of Golden Rice: A Philippine case study. *Food Policy, 29*(2), 147–168. https://doi.org/10.1016/j.foodpol.2004.03.001

Zyba, S. J., Shenvi, S. V., Killilea, D. W., Holland, T. C., Kim, E., Moy, A., et al. (2017). A moderate increase in dietary zinc reduces DNA strand breaks in leukocytes and alters plasma proteins without changing plasma zinc concentrations. *The American Journal of Clinical Nutrition, 105*(2), 343–351. https://doi.org/10.3945/ajcn.116.135327

Open Access This chapter is licensed under the terms of the Creative Commons Attribution 4.0 International License (http://creativecommons.org/licenses/by/4.0/), which permits use, sharing, adaptation, distribution and reproduction in any medium or format, as long as you give appropriate credit to the original author(s) and the source, provide a link to the Creative Commons license and indicate if changes were made.

The images or other third party material in this chapter are included in the chapter's Creative Commons license, unless indicated otherwise in a credit line to the material. If material is not included in the chapter's Creative Commons license and your intended use is not permitted by statutory regulation or exceeds the permitted use, you will need to obtain permission directly from the copyright holder.

Agronomic Biofortification of Crops with Zinc: A Comprehensive Overview

Raheela Rehman, Muhammad Moaz Latif, Muhammad Ahsan Khan, and Zaheer Ahmed

Introduction

In many parts of the world, people rely on cereals as a staple food source and cereals are low in Zn concentration which is one of the leading causes of malnutrition. Zinc (Zn) is an essential micronutrient for plants and its deficiency in soil leads to stunted growth, underdevelopment, diminished yield, and scarce nutritional profile of crops (Kadi et al., 2018). The World Health Organization (WHO) assessment is that approximately 31% of the people in the world and 4–5% in specific parts of the globe are afflicted by zinc paucity, accentuating the necessity to address the challenge. 60–70% of South Asian people are encountering zinc shortage (Gibson, 2006), and it is the cardinal reason of diminutive growth in children globally. In Pakistan, 12 million children are stunted, while 22.1% of women and 18.6% of under-five children are low in zinc. Punjab has the highest zinc deficiency rate in women (24.1%), followed by Balochistan (23.4%) and Sindh (21.4%), while

---

R. Rehman (✉)
Department of Plant Breeding and Genetics, University of Agriculture Faisalabad, Faisalabad, Pakistan

Pakistan-Korea Nutrition Center, University of Agriculture Faisalabad, Faisalabad, Pakistan
e-mail: raheela.rehman@uaf.edu.pk

M. M. Latif · Z. Ahmed
Department of Plant Breeding and Genetics, University of Agriculture Faisalabad, Faisalabad, Pakistan

Center for Advanced Studies in Agriculture and Food Security, University of Agriculture Faisalabad, Faisalabad, Pakistan
e-mail: 2017ag9809@uaf.edu.pk; zaheer.ahmed@uaf.edu.pk

M. A. Khan
Department of Plant Breeding and Genetics, University of Agriculture Faisalabad, Faisalabad, Pakistan
e-mail: ahsankhanpbg@uaf.edu.pk

© The Author(s) 2025
M. Govindaraj et al. (eds.), *Breeding Zinc Crops for Better Human Health*,
https://doi.org/10.1007/978-3-031-84342-6_8

Khyber Pakhtunkhwa has the lowest frequency (15.9%) (National Nutrition Survey 2018 | UNICEF Pakistan, n.d.).

Zinc scarcity is a prevailing crisis in most regions of the world, especially in areas with alkaline or sorely weathered soils. Such inadequacies have detrimental effects on plant vigor and human sustenance because zinc is necessary for human growth, immunological activity, and cognitive development (Brown et al., 2001). Most of the soil under cultivation in the world has low Zn plant availability due to which Zn biofortified genotypes are not able to perform at the potential of the required Zn concentration in grain (Cakmak & Kutman, 2018). These outcomes manifest that over 50% of the population of the world is in danger of zinc deficiency, underlining the significance of immediate public health efforts to solve the issue of zinc shortage. The unavailability of biomarkers for zinc status has constrained the global estimates of the occurrence of zinc deprivation.

Conventional, molecular, genetic (transgenic) and agronomic interventions are major tools for Zn biofortification, a method of improving the zinc content of edible crops. Among the numerous techniques used, zinc soil, foliar, and seed treatments are crucial, sustainable, feasible and cost-effective approaches to enhance the Zn content of wheat, rice, maize, etc. (Hussain et al., 2012). Crop productivity is limited by the soil Zn deficiency, therefore, the availability of adequate Zn to plant is necessary and there is a demand to explore complementary techniques for its management, such as soil, foliar, and seed treatments.

Soil, Foliar, and Seed Application of Zinc for Biofortification

Soil Application of Zn for Biofortification

Applying zinc to the soil for biofortification is a reproducible method because it harmonizes with the existing agricultural approaches and has minimal hazardous impacts on the environment (Alloway, 2008). Ample Zn availability in soil stimulates vigorous crop growth and enhances yield and agricultural productivity as a whole (Cakmak, 2002). Soil implementation of zinc increases zinc concentrations in edible plant parts, which contributes to crop biofortification and addresses zinc insufficiency in human diets (White & Broadley, 2009). Zinc fertilization had a negligible effect on rice grain yield, except for increases, on average, Zn treatment raised grain yield by roughly 5%, but in rare cases up to 10%. However, Zn fertilization was more successful in increasing grain Zn concentrations (Phattarakul et al., 2012). Soil application is a method of augmenting soil with zinc-containing fertilizers or transformants. This attempts to improve the zinc concentration in crops, hence augmenting their nutritional significance. Numerous soil application strategies are employed to amend the accessibility and absorption of zinc by crops.

Zinc Fertilizers

Zinc-containing fertilizers are a frequent and effective way to provide zinc to soils. The application of Zn fertilizers is one of the successful ways to combat Zn shortage. A broad range of Zn sources are available on the market, however, the most widely used fertilizers are ZnO, $ZnSO_4$, zinc chelates, or other soluble zinc compounds that plants may absorb. The chemical relation between soil and zinc controls the availability of zinc. These interactions are influenced by the method of fertilizer application and the chemical and physical characteristics of the fertilizer, for instance, granule size, the physical state of solid versus fluid, and the water solubility of zinc (Alloway, 2008; Montalvo et al., 2016). Broadcasting, banding, or foliar application methods are used to deploy the Zn-containing fertilizers. Soil amalgamation guarantees proper and homogeneous zinc dispersion throughout the root zone. Zinc fertilizers improve the crop assimilation of zinc and resolve deficits of zinc. Farmers fertilize the soil with zinc-containing products before planting or throughout the crop growth and development.

Zinc-Enriched Organic Amendments

Integrating organic alterations enhanced with zinc is an expedient technique. Some regions may have soils that are scarce in zinc, which necessitates the need for certain modifications tailored to elevate the amount of zinc in soil. These combinations may include treating soil with organic matter, manure, compost, or other products supplemented with Zn. Organic substances i.e. compost/manure can be augmented with compounds rich in zinc. Introducing organic modifications enhanced with zinc can ameliorate soil fertility and elevate crop zinc uptake by enhancing Zn availability (Cakmak, 2008; Impa & Johnson-Beebout, 2012). These interventions not only contribute to the soil organic matter but also advance the availability of zinc for the plants to absorb. This approach has been established as dependable and valuable in promoting sustainable agriculture.

Precision Agriculture Technologies

Advanced technology, like precision agricultural techniques, can be used to target zinc applications. These methods aid in improving zinc treatment by targeting parts of the field when a zinc shortage is found by soil testing (White & Broadley, 2009). These practices include assessment of soil to comprehend zinc deficits, and permitting meticulous application of zinc fertilizers at the required time. This approach optimizes the consumption of resources and mitigates the impacts on the environment (Smith et al., 2013).

Zinc-Containing Irrigation Water

Water irrigation with dissolved zinc can prove to be favorable for territories with soils low in Zn. This promises the sustainable provision of zinc to crop plants during the wheat growing season. In comparison to the control zinc conditions, zinc accumulation in plant tissues improved significantly under all treatments, with dose-dependent rises in roots, shoots, and leaves Zn concentrations (Reshma & Meenal, 2024). However, the applicability of this approach is based on water quality and potential interactions with other soil elements.

Each of these methods has its advantages and considerations. The application technique is decided by soil conditions, crop type, and the severity of zinc shortage. Combining different techniques may be needed for complete zinc control in agricultural systems. Zinc biofortification by soil application is a direct and long-term method of increasing the food intake of this critical mineral. Soil application methods are often less expensive than other biofortification approaches, making them affordable to small-scale farmers. By improving soil health and fertility, zinc treatment promotes sustainable agriculture by fostering crop yield and enhancing soil quality. Efficacious soil application techniques for zinc biofortification require a comprehensive strategy that takes into consideration the target population's nutritional well-being as well as crop types, agricultural practices, and soil conditions.

Foliar Application of Zn for Biofortification

Foliar application of zinc for biofortification entails spraying a zinc-containing solution directly onto plant leaves. This approach circumvents any soil-related constraints on zinc uptake and enables quick zinc absorption by plant leaves. Zinc sulfate or other soluble zinc compounds are dissolved in water and sprayed onto the plant's leaves. Afterward, the plant absorbs the zinc through the stomata on the leaf surface, and it is then translocated to a variety of parts of the plant, including edible tissues. Foliar therapy is advantageous since it is an instantaneous method to fix crop zinc deficiencies when soil zinc availability is limited. Foliar treatment is also efficacious in the provision of zinc directly to the plant. However, to sustain adequate zinc ranges in the plants, a number of Zn treatments can be required to apply for the crop growing season. Additionally, the time and concentration of the foliar spray are important for successful biofortification.

Through foliar zinc treatment, an efficacious and focused biofortification method, zinc is directly applied and delivered to plant leaves (Marschner, 2011). When crop absorption of zinc is impeded by soil conditions, this approach proves to be particularly favorable. Foliar application is the process of spraying a chemical solution containing zinc onto leaves, facilitating rapid stomata absorption. The effectiveness of foliar therapy is highly dependent on the selection of zinc components and the growth stage of plants at which it is sprayed. The solubility and absorption efficacy are influenced by the selection of zinc compounds (Alloway, 2008).

In different crops, the foliar zinc administration approach has emerged as efficient in addressing zinc deficiencies. For instance, foliar zinc therapy on wheat (*Triticum aestivum* L.) has demonstrated positive effects, especially at crucial developmental stages such as flower initiation and grain filling (Cakmak, 2008). When zinc is applied through foliar spray on maize (*Zea mays*), particularly grown on soils lacking in zinc, it has shown increased yield and enhanced zinc levels in kernels. (Anwar et al., 2021). Similarly, rice (*Oryza sativa*) zinc treatment through foliar spray improves the zinc content of the grain and alleviates its nutritional status (Widodo et al., 2010). Moreover, Zinc deficiencies in soybean (*Glycine max*) crop has also been addressed using the foliar application method of zinc, which resulted in improved plant development and yield (Alloway, 2008).

Furthermore, the foliar spray application method of zinc helps biofortify fruits, particularly citrus trees such as oranges and lemons (Citrus spp.) demonstrated increased Zn concentrations in fruit tissues (Bhantana et al., 2022). Foliar zinc spray's versatility and effectiveness in raising zinc levels and nutritional quality across a wide range of plant species are demonstrated by the wide range of crops.

Foliar spraying, as opposed to soil treatment, reduces the overall environmental impact by allowing plants to absorb zinc more rapidly and correcting zinc shortages promptly by focusing primarily on the affected plants (Marschner, 2011). Foliar application of zinc at critical growth stages brings flexibility in rectifying zinc deficiencies during specific phases of crop growth and development. Foliar treatment is highly efficacious when applied during periods of vigorous plant development and high zinc demand. The efficiency of foliar spray depends on ensuring consistent foliage coverage. Proper spraying procedures and equipment are key factors (Hong et al., 2021). Zinc applied using foliar spray may have a lower residual impact than zinc applied to the soil. Various applications may be necessary throughout the growth season. Some crops may be more susceptible to foliar treatment thus vigilance is essential to avoid phytotoxicity (Niu et al., 2021).

Seed Application of Zn for Biofortification

Applying zinc to the seed is a strategic biofortification technique that guarantees zinc availability to growing crop plants at an early stage. This technique involves coating seeds with formulations containing zinc prior to sowing, which offers a targeted and effective delivery of zinc throughout the critical early plant growth stages (Maqbool & Beshir, 2019). Applying zinc to seeds guarantees an accessible pathway for Zn to the developing seedling, and encourages the growth of early roots and shoots (Rengel, 2001). Zinc coating of seeds endorses homogeneous dispersion of nutrients to the plant parts, supporting continuous plant growth and development throughout the field (Alloway, 2008). This consistent distribution is crucial in hampering confined nutrient deficiencies and guarding the continuous access to adequate zinc at the early development phases for all plants. Targeted transport of zinc directly to seedlings minimizes nutrient waste to the surrounding soil, limiting overall environmental effects in comparison to substitute methods of treatment (Mengel et al., 2001). This meticulous nutrient distribution reduces runoff and probable environmental adulteration.

Zinc formulations such as in the form of sulphate or oxide coating can be applied to seeds. The choice is made depending on factors like the type of seed and the state of the soil. Recent advances in seed-coating technology have enabled the creation of more effective and sustainable zinc compositions. The application procedure is critical to the success of seed coating. Film-coating or encrusting techniques are used to provide a homogeneous and adherent zinc covering on the seed surface (Alloway, 2009). To enhance the transport of zinc to seedlings, several approaches to coat seeds have been studied, especially applications of nanotechnology in seed coating.

To prevent any probable antagonistic effects, cross-talks between the signaling pathways of zinc and other nutrients, such as phosphorus (P) and iron (Fe), should be investigated. In order to reduce nutrient imbalances and maximize its use efficacy, contemporary research has explored the model nutrient ratios and formulations for seed coatings (Mengel et al., 2001). Similarly, the sustenance of the zinc coating's efficacy throughout germination and the early stages of development depends on the viability of the seed coat during storage and seed sowing. Advancement in seed-coating technology, especially the development of biodegradable coatings, resolves the issue of seed coat robustness and environmental sustainability (Rengel et al., 1999). To ameliorate overall plant health, zinc seed coating should be incorporated with supplementary seed treatments, such as fungicides or insecticides. Proving integrated seed-coating systems with several nutrients and agronomic inputs for inclusive seed application has been the focus of modern research (Gao et al., 2023).

Zinc Sources for Biofortification: Zinc Salts Versus Zinc Chelates

The two main methods for increasing crop zinc levels through biofortification are zinc salts and zinc chelates; however, their chemical compositions and plant-uptake capabilities are different.

Zinc Salts

More than 2 billion people suffer from micronutrient malnutrition, which is mostly caused by zinc deficiency in soils and a poor human diet. To address this, research recommends adding zinc to major food crops. Agronomic biofortification, which employs Zn-based fertilizers, is both affordable and simple to implement. Although time-consuming, genetic biofortification is remarkably successful and may be used for many years without incurring further expenditures. Both of these approaches can be efficacious in treating zinc insufficiency (Praharaj et al., 2021).

In damp soil, inorganic compounds known as zinc salts break down easily and release zinc ions. These ions are easily accessible to plant roots for uptake, which may be absorbed and integrated into its tissues. By providing easily absorbed forms of zinc, these water-soluble chemicals affect plant absorption and treat zinc deficiency. There are many different chemical forms of zinc salts. Three prominent ones are zinc sulfate ($ZnSO_4$), zinc nitrate ($Zn(NO_3)_2$), and zinc chloride ($ZnCl_2$) (Alloway, 2008). Every salt has distinct qualities that affect its handling, solubility, and plant absorption capacity. In agriculture, Zinc sulphate monohydrate ($ZnSO_4 \cdot H_2O$) is the most widely applied water-soluble zinc salt. It can be foliar sprayed on leaves or added straight to the soil to provide plants with a robust source of zinc for uptake.

Plant availability of zinc salts is significantly influenced by their solubility. The likelihood of root absorption is increased by highly soluble forms, like zinc sulphate, which quickly releases zinc ions into the soil solution. This characteristic is crucial in determining the efficiency of zinc salts in supplying crops with essential nutrients. There are various strategies to apply zinc salts, such as foliar spray and soil application. Even though zinc salts are effective in biofortification, Zinc availability can be influenced by the interactions with elements like phosphorus and iron, (Cakmak & Kutman, 2018). Moreover, considerable research is needed to understand the pH dependence of zinc salts, especially in soils with high pH. These factors alter the overall efficiency of zinc salts in crop biofortification.

Zinc Chelates

Zinc chelates appear as a key source of biofortification, differing from zinc salts in that they have a distinctive chemical form and are readily available to plants. Chelates are organic compounds that encircle and encapsulate metal ions like zinc, resulting in a stable and soluble complex. Zinc ions bound to ligands, which are organic molecules, create stable complexes called zinc chelates. By preventing zinc from binding with other soil constituents, these chelates help improve plants' zinc accessibility.

EDTA and DTPA

EDTA (ethylenediaminetetraacetic acid) and DTPA (diethylenetriaminepentaacetic acid) are two examples of chelating chemicals utilized in a variety of techniques and approaches. Plants receive a controlled and extended percentage of zinc from EDTA and DPTA, which gradually improves the bioavailability of Zn. One chelating compound that is frequently used in biofortification is EDTA. With zinc, it forms a stable compound that improves solubility and bioavailability for plant uptake (Arts et al., 2018). EDTA is a synthetic compound and this property distinguishes it from the rest of the natural chelates.

Amino Acids and Proteins

The production of ligands by proteins and amino acids like histidine and glycine in nature is essential for the translocation of zinc to plants. These chemical chelates, which are organic in origin, stabilize zinc ions, increasing the efficiency of plant roots' absorption of zinc. Zinc chelates, like zinc salts, are delivered by soil and foliar techniques. Chelates are applied to the soil directly or through chelate-containing fertilizers. The foliar application involves spraying zinc chelates onto plant foliage to ensure fast absorption by leaves.

Zn chelates increase the availability of zinc by inhibiting precipitation or fixation in the soil, easing higher absorption by plants (White & Broadley, 2009). Chelates provide benefits in instances where soil conditions might normally limit zinc availability. The choice between zinc salts and zinc chelates for biofortification is decided by soil conditions, crop needs, environmental conditions, and preferred manner of application (Almubarak et al., 2021). Zinc salts are inexpensive and commonly used, particularly in soils with sufficient organic matter, but zinc chelates are helpful in alkaline or calcareous soils where zinc availability is often limited. For example, zinc sulfate might be compared to zinc EDTA. Zinc sulfate provides a quick but short supply of zinc, while zinc EDTA, as a chelated form, provides a more stable and prolonged release of zinc, perhaps providing greater availability for plant uptake over time (Suganya et al., 2020). Zinc chelates increase zinc availability and absorption efficiency. These chelates' organic quality supports zinc ion stability and reduces the likelihood of their interaction with other micronutrients. However, it is necessary to consider the costs incurred as a result of these ligands and their inherent environmental impacts.

Zinc salts and chelates can be applied efficiently to address zinc deficiencies in crops depending on the type of plant, soil conditions, method of application, and desired period of zinc availability. Comprehending the characteristics and interactions of various sources of zinc is essential for developing efficacious biofortification programs customized for specific farming environments.

Agronomic Biofortification of Cereal, Legumes, Vegetables, Forage, and Fodder Crops with Zinc

To enhance the nutritional content of cereal crops including wheat, rice, maize, barley, sorghum, millet, triticale, rye, buckwheat, and other pseudocereals requires the use of agronomic biofortification methods. It can improve public health outcomes worldwide and combat malnutrition by raising the levels of important micronutrients like zinc. Agronomic biofortification provides a solution without sacrificing yield or product acceptance issues. It is figured out by management strategies, soil conditions, and plant variables. Genetic and agronomic biofortification work together, but more study is needed to understand the complicated soil-plant-management relationship under various agroecological circumstances, notably for rice (Prasad et al., 2014).

Agronomic Biofortification of Cereal

Wheat

It is important for the global food chain as a staple crop that provides nourishment for millions of people. Along with other techniques, foliar application and soil fertilization are used in agronomic biofortification strategies for wheat in an attempt to improve the quantity of nutrients it contains, particularly zinc. Soil fertilization is a crucial and major technique for the agronomic biofortification of wheat. Research has indicated that zinc sulphate is an effective micronutrient fertilizer for considerably increasing zinc contents in wheat grains (Stomph et al., 2009). In addition, current studies have explored novel fertilizer compositions and delivery techniques with the goal of improving zinc absorption and translocation in wheat plants (Liu et al., 2019). Technological developments in soil fertilization have aided in the creation of wheat varieties that are biofortified and have higher nutritional values, particularly in zinc content. Applying foliar micronutrient solutions directly to wheat leaves is another efficient method for zinc agronomic biofortification of wheat. The usefulness of foliar sprays in boosting iron and zinc concentrations in wheat grains has been highlighted by recent research, especially at critical stages of development (Ullah et al., 2022). Additionally, studies have looked at the foliar application of chelated micronutrient formulations, which improve wheat plants' absorption and utilization of zinc (Raza et al., 2017). Techniques for foliar spraying provide a focused way to increase wheat grains' nutritional value while reducing the negative environmental effects of overuse of fertilizers. These outcomes demonstrate the possibility of foliar spray as a workable and effective way to improve wheat crop zinc biofortification.

Rice

Rice is a staple food for a substantial part of the world's population, especially in places where it is the predominant source of nutritional energy. Agronomic biofortification approaches for rice, especially zinc enrichment, have appeared as practical solutions to treat zinc shortage and enhance human health outcomes. Genetic engineering and conventional plant breeding are being used to help address the iron and zinc deficits that exist in emerging Asian and African nations. However, there is a trade-off between yield and grain biofortification, and these techniques are not commonly used. Agronomic biofortification provides a solution without compromising production or creating problems with product acceptability. It is dependent upon plant, soil, and management approaches (Prasad et al., 2014). Although agronomic and genetic biofortification are complementary, more study is needed, especially for rice, to fully understand the intricate interactions between soil, plants, and management in various agroecological conditions.

An important and initial step in the agronomic biofortification of rice with zinc is soil fertilization. Micronutrient fertilizers, including zinc sulphate, have been demonstrated to significantly raise zinc content in rice grains (Rakotondramanana et al., 2024; Widodo et al., 2010). These results suggest the importance of maximizing soil bioavailability of zinc in order to increase zinc intake ultimately enhancing the nutritional value of rice. Furthermore, novel fertilizer compositions and techniques for application are being investigated to perfect nutrient uptake and translocation in rice plants (Liu et al., 2024). Research in this area continues to evolve, aiming to develop biofortified rice varieties with enhanced zinc content suitable for addressing global malnutrition.

Applying foliar micronutrient solutions directly to rice leaves is another efficient method for zinc agronomic biofortification of rice. Increasing zinc concentrations in rice grains using foliar sprays has proven encouraging outcomes in recent research, especially at essential development phases (Zulfiqar et al., 2021). Furthermore, research has investigated the use of chelated micronutrient formulations for foliar application, which promote zinc absorption and use by rice plants (Impa & Johnson-Beebout, 2012). Application by foliar spray technique offers a highly specific and effective strategy for improving zinc contents in rice crops while keeping the environmental harms associated with fertilizer application to a minimal level.

Maize

Maize, an important cereal crop, is a source of staple food for humans and an essential component of animal feed. Agronomic biofortification techniques for maize, notably focused on zinc enrichment, have received attention to alleviate zinc shortage and improve human health outcomes.

Soil fertilization is the principal strategy in agronomic biofortification of maize with zinc. Recent studies have shown that micronutrient fertilizers, such as zinc sulphate, may dramatically increase zinc content in maize grains (Cakmak & Kutman, 2018; Impa & Johnson-Beebout, 2012). These results highlight how crucial it is to maximize soil zinc availability to boost maize's nutritional value by increasing zinc absorption. Furthermore, novel fertilizer compositions and techniques of administration are being investigated to perfect nitrogen uptake and translocation in maize plants (Gao et al., 2023). The goal of this field's ongoing research is to create biofortified maize cultivars with higher zinc contents that may be used for fighting worldwide hunger.

Another potential technique for zinc agronomic biofortification of maize is foliar spraying of micronutrient solutions directly to the maize leaf. Recent research has proved encouraging results in raising zinc contents in maize grains via foliar sprays, particularly during crucial development periods (Mehboob et al., 2022). Furthermore, research has studied the use of chelated micronutrient formulations for foliar application to promote zinc absorption and utilization by maize plants (Eifediyi et al., 2021). Application by foliar spray technique offers a highly specific and effective strategy for improving zinc contents in maize with the least amount of environmental risk associated with fertilizer application.

Other Cereals

In addition to wheat, rice, and maize, other cereals such as barley, sorghum, millet, oats, rye, and triticale are essential crops that are grown globally, adding a significant quantity to address the challenge of global food security. Agronomic biofortification, targets to improve the nutritional quantity of crops, especially through zinc enhancement, has emerged as an indispensable approach in combating micronutrient deficiencies and augmenting human well-being.

Soil Fertilization Soil fertilization is the principal method for agronomically biofortifying these cereal crops with zinc. Recent studies have proved the usefulness of micronutrient fertilizers, such as zinc sulphate, in considerably raising zinc levels in grains (Impa & Johnson-Beebout, 2012).

To improve zinc intake and nutritional quality of cereal crops, soil bioavailability of zinc must be maximized. To optimize nitrogen uptake and translocation within plants, novel fertilizer compositions and techniques to apply it are being explored (Rizwan et al., 2021). The creation of biofortified grain varieties with enhanced zinc concentrations, that can be utilized to address hidden hunger, is the objective of the ongoing study.

Foliar Application Foliar spraying of micronutrient solutions to these cereal crops' leaves is another efficient technique for zinc agronomic biofortification. Recent research has revealed encouraging results in raising zinc concentrations in grains via foliar sprays, particularly during crucial development periods (Ullah et al., 2022). Use of chelated micronutrient formulations for foliar application has been shown to enhance zinc uptake and use by plants (Hassan et al., 2024). Foliar application techniques offer a targeted and efficient method for enhancing zinc biofortification in cereal crops while minimizing environmental impacts associated with excessive fertilizer use.

Agronomic Biofortification of Legumes

Legumes contain a wide range of types of plants including peas, beans (soybean, Faba bean, etc), lentils, and chickpeas. These plants provide vital nutrients and, therefore, are used as staples in diets worldwide. Methodologies to biofortify legumes, especially those that are focused on increasing zinc levels in the edible parts of the legume crops, are crucial in addressing hidden hunger or malnutrition.

An important strategy for the agronomic biofortification of legumes with zinc is soil fertilization. Research activities have proved that using fertilizers having zinc, like zinc sulfate, may greatly raise the zinc contents in legume grains (Bana et al., 2022). To improve legume absorption of zinc and hence their nutritional value, soil zinc availability must be maximized. To enhance nutrient absorption and translocation inside bean plants, novel fertilizer formulations and delivery techniques are being investigated (Ali et al., 2022).

Soil Fertilization Applying micronutrients to the soil through fertilization technique is a major strategy in agronomic biofortification with zinc. Modern research endeavors have demonstrated the positive impacts of micronutrient fertilizers, particularly zinc sulphate, which can significantly raise the zinc contents in legume seeds (Athar et al., 2020). Improving the availability of zinc in soils is essential in enhancing Zn uptake by legume crop plants, ultimately increasing its nutritional value. To optimize the uptake and transmission of nutrients inside plants, novel fertilizer chemical compositions, and application strategies are being explored (Stanton et al., 2022). In this field, the research in recent times is aimed at developing new cultivars that have been biofortified for high zinc concentration to support fight nutritional deficiency.

Foliar Application Foliar spraying of micronutrient solutions to legume leaves is another successful technique for zinc agronomic biofortification. Recent studies have revealed encouraging results in boosting zinc concentrations in legume grains via foliar sprays, particularly during crucial development periods Utilization of chelated micronutrient formulations for foliar application has been shown to enhance zinc uptake and use by legume plants (Yeboah et al., 2021). Techniques for foliar spraying provide a focused and effective way to maximize zinc biofortification in legume crops while reducing the negative environmental effects of over-fertilizer use.

Zinc agronomic biofortification may also be achieved by foliar spraying micronutrient solutions directly onto the leaves of legume crops. Increasing zinc concentrations in legume seeds using foliar sprays has shown encouraging outcomes in recent research, especially during crucial development periods (dos Santos et al., 2022). Legume plants have been proven to absorb and use zinc more efficiently when chelated micronutrient formulations are applied topically (Stanton et al., 2022). Legume plants have been proven to absorb and use zinc more efficiently when chelated micronutrient formulations are applied directly.

Agronomic Biofortification Vegetables

Vegetables are an important source of nutrients for humans and therefore are used in the diet by almost everyone in the world. The techniques such as soil application and foliar spray are principal methods of biofortification that ameliorate the micronutrient percentage in vegetables and other crops, with a major focus on the enhancement of zinc. Therefore, agronomic enrichment of zinc and other micronutrients is crucial in resolving nutritional shortages and improving human health.

Spinach

Spinach is rich in iron (Fe) and its edible part is green leaves. Agronomic techniques developed for the enrichment of micronutrients are directed towards improving iron bioavailability for spinach. Iron biofortification has been achieved through soil application of fertilizers and foliar sprays of containing iron (Adnan et al., 2021).

Carrots

Carrots are one of the widely used vegetables and are rich in beta-carotene. In carrots, soil amendments and breeding programs have been established to biofortify them with a precursor of vitamin A i.e. beta-carotene. Fertilizers containing boron (B) and zinc (Zn) are applied to the soil to achieve this goal agronomically (Dimka Haytova, 2013).

Tomatoes

Tomatoes are consumed by humans either fresh or cooked and these are a source of antioxidants i.e. lycopene. Agronomic and Genetic biofortification studies have been carried out to explore the possibility of concentrating lycopene into tomatoes. Several amendments to the soil have been successfully performed by treating it with manganese and copper (Krishnasree et al., 2021).

Agronomic Biofortification of Fodder and Forages

Forages and fodder are vital parts of the animal diet and provide essential nutrients for the growth and development of livestock. To prevent animal health risks and nutrient deficits, agronomic biofortification strategies should be adopted to improve the micronutrient concentration of fodder and forage crops, especially for zinc embellishment.

Soil Fertilization Zinc and iron deficits in developing countries are associated with low-quality staple food grains, which contribute to human malnutrition. To address these inadequacies, diet diversity, food fortification, dietary supplements, and biofortification are being emphasized. Cereal crops are the primary goal, although pulses are the second-best alternative owing to their nutritious content. Micronutrient fertilization of oilseed crops is critical for increasing edible oil output and nutritional quality. Fruits and vegetables can receive help from zinc and iron supplementation, and future micronutrient biofortification should consider soil-plant-animal-human interactions. Market incentives and organic product pricing strategies are also critical to success (Shahane & Shivay, 2022). Optimizing zinc availability in soils is critical for increasing zinc absorption by fodder and forage crops, which improves their nutritional value. Innovative fertilizer formulations and administration strategies are being investigated to maximize nutrient uptake and translocation inside plants.

Foliar Application Micronutrient deficits in fodder crops in underdeveloped nations impact animal health and production. Biofortification is a long-term method for reducing these inadequacies. Studies have shown that foliar treatment can increase copper and zinc concentrations in animal feed. Agronomic biofortification is a quick and cost-effective method for at once enriching feed and alleviating animal malnutrition (Singh Dhaliwal et al., 2023).

Another effective methodology to biofortify zinc is the foliar spray i.e. applying solutions of zinc and other micronutrient-containing compounds directly to the fodder and forage leaves. In modern studies, this technique of direct application has uncovered promising outcomes in raising zinc contents in the tissues of plants, especially when the plant is at critical growth stages (Ali et al., 2022). Foliar application methods demonstrate a targeted and efficacious approach to ameliorating zinc in these crops while minimizing the environmental challenges associated with excessive fertilizer use.

Combining Zinc with Other Minerals for Simultaneous Biofortification

Using a variety of essential minerals, simultaneous biofortification offers a comprehensive approach to correcting nutritional deficiencies in staple crops. When zinc is combined with other macro- and micronutrients, crop nutritional quality is enhanced, contributing to a more inclusive approach to the issue of malnutrition. The results of the research shown below demonstrate the importance of addressing zinc interactions with other minerals in order to optimize crop yield, nutrient uptake, and plant development.

Zinc and Nitrogen (N)

Even though nitrogen is a major macronutrient, its interaction with zinc is crucial for plant growth, health, and development. The combined effects of zinc and nitrogen on crop development prove the two elements' synergistic interaction in enhancing grain yield and nutritional quality. During the reproductive period, applying zinc and nitrogen, iron and nitrogen, or zinc, iron, and nitrogen to the soil or foliage results in increased nutritional (zinc and iron) content in the plant's edible sections. This emerges due to the reason that remobilization and transportation of zinc, iron, and urea follow the same genetic and biochemical pathways inside the plant (Kaur & Singh, 2022).

Zinc and Phosphorus (P)

Phosphorus is an indispensable macronutrient that is essential for the transmission of energy and many metabolic functions. Research (Bibi et al., 2020) investigated the interactions between zinc and phosphorus on maize growth and nutrient absorption. It also reduces the phytic acid concentration and improves the availability of Zn consumption in humans. The study found that applying zinc and phosphorus together improved root shape, phosphorus usage efficiency, and grain production in maize crops, suggesting a synergistic link between the elements.

Zinc and Potassium (K)

Potassium is another vital macronutrient that plays an important function in osmoregulation and enzyme activity and potassium (K^+) interacts with other nutrients to boost plant potential against abiotic stress which induces negative effects on plants. A study (Mostofa et al., 2022) on the combined effects of zinc and potassium on rice plants. The study found that combining zinc and potassium boosted nutrient absorption, biomass formation, and grain production in rice, emphasizing the potential benefits of improving the zinc-potassium interaction.

Zinc and Iron

Deficiency in zinc and iron frequently coexist, endangering human life. Agronomic research, that has already been conducted, demonstrates the success potential of applying zinc and iron simultaneously to the soil, a viable method of biofortification. Fertilizers that contain zinc when strengthened with iron help plants absorb these elements better, leading to improved cereal grain quality (Zulfiqar et al., 2020). There has been an emphasis on applying zinc and iron simultaneously due to the coordinated occurrence of deficiency in both elements. The concurrent application of zinc and iron salts or chelates exceptionally improved the concentrations of both elements in wheat grains, resolving deficits in areas with calcareous soils (Cakmak & Kutman, 2018).

Zinc and Selenium

Another necessary component is selenium, which has proven to be beneficial when applied in combination with zinc to the soil. Zinc and Selenium work synergistically to improve the nutritional content for crop biofortification, especially in areas where selenium is scarce (Rayman, 2012). The health effects of selenium are intrinsically linked to status; while increased selenium consumption may help people with low status, those with adequate-to-high status may be negatively affected and should avoid selenium supplementation.

Zinc and Manganese (Mn)

Alloway (2008) emphasizes the significance of understanding the relationship of zinc and manganese in the soil. Although both micronutrients are essential, the presence of one in excess could limit the percentage of the other. To sustain sufficient nutritional balance in plants and reduce antagonistic interactions, it is imperative to maintain an optimal zinc-to-manganese ratio.

Zinc and Copper (Cu)

Notably, plants may benefit from the combined effects of copper and zinc on their growth and development. In comparison to applying nutrient treatments solely, the coordinated application of zinc and copper improves grain yield by a higher percentage in barley and wheat, underlining the significance of a balanced micronutrient provision (Cakmak, 2002).

Zinc and Boron (B)

Although boron and zinc are both necessary for several physiological processes, their coordinated effects have been demonstrated to improve crop quality. Applying zinc in combination with boron improved the size, color, and nutritional content of apple tree fruit (Marschner, 2011).

Zinc and Magnesium (Mg)

The enzymatic activity of plants is considerably influenced by the interactions between zinc and magnesium. The coordinated zinc-magnesium effects on the growth, development, and nutritional profile of maize were examined in a recent study (Li et al., 2020). The investigation accentuated that the zinc and magnesium crosstalk is in concert with the betterment of plant growth and nutrient absorption.

Zinc and Calcium (Ca)

Although calcium and zinc serve different functions in plants, both these elements must be applied in balance. Research findings in barley indicated that the coordinated supplementation of calcium and zinc enhanced root elongation, this demonstrates that it is highly indispensable to provide balanced nutrients (Rengel et al., 1999).

Zinc and Molybdenum (Mo)

Molybdenum plays a crucial role in the metabolism of nitrogen in plants, therefore, its interaction with zinc is considered vitally important for ameliorating the growth and development of cereals and other crops. For instance, in an experiment on chickpea crops, the application of molybdenum in combination with zinc marked betterment in yield and advancement in the absorption of nitrogen (Singh Dhaliwal et al., 2023).

Zinc and Sulfur (S)

Sulfur is involved in the vital process of protein synthesis in plants, and the interaction of this micronutrient with zinc regulates the growth of the crop plant. In a study conducted to augment the absorption of nutrients and improve the quality of seed, it was found that the combined application of zinc and sulfur has a positive effect on the growth of soybean plants (Zhang et al., 2016).

The application of zinc in combination with other macro and micronutrients has varying impacts on the growth of plants, nutrient uptake, and crop nutritional value. These crosstalks between the micronutrient pathways are convoluted and context-dependent, with the condition of the soil, crop genetics, and nutrient percentage all having an impact. Understanding these relationships is critical for creating efficient nutrient management techniques that maximize plant nutrition and increase agricultural sustainability. While simultaneous biofortification has promise, it is not without obstacles. Nutrient interactions, soil properties, and crop-specific responses must be carefully considered. Balancing zinc and other mineral ratios are critical for avoiding antagonistic effects and ensuring proper nutrient uptake by plants. Future study should concentrate on improving zinc combinations with other minerals while considering soil conditions, crop kinds, and nutritional requirements. Innovative technologies, such as precision agriculture, may help to customize simultaneous biofortification tactics to individual agroecosystems.

Effects of Zinc Applications for Biofortification on the Concentrations of Iron and Other Minerals

Zinc (Zn) biofortification can change the amounts of other minerals in plants. Zinc's interactions with other minerals, notably iron, are complicated and context-dependent. The kind and amount of zinc used, as well as other agronomic methods, can affect plants' overall response to biofortification and the effects on mineral concentrations. It is crucial to note that, while zinc biofortification intends to improve zinc concentrations in crops, the possible effects on other minerals highlight the necessity for a comprehensive approach to nutrient management. Zinc uptake in plants is a complex process that begins with absorption via root transporters. This mechanism affects a variety of physiological activities, including enzyme activation and hormone control. Zinc transport through xylem and phloem contributes to its overall impact on nutrient concentrations in the plant. Soil conditions, plant species, and agronomic techniques must be considered when maximizing biofortification procedures and avoiding unanticipated negative impacts on plant nutrition. Zinc biofortification has varying effects on other mineral concentrations in various plant species. Some plants may react more strongly to variations in zinc levels, while others may be less impacted. Field trials and research spanning diverse settings and crops are needed for a thorough understanding of these connections.

Effect of Zn Application on Fe Concentration

The effect of biofortification of zinc (Zn) on the concentration of iron (Fe) is of essential discussion, particularly in the context of resolving nutrient deficiencies. The interaction between zinc and iron is complex, including both direct and indirect linkages. Based on the extant research, the following are some critical factors to consider. Zinc and iron have similar intake routes in plant roots. When zinc is taken in higher amounts, it may competitively hamper iron absorption by the roots. Rengel et al. (1999) studied barley plants and found that increasing zinc levels resulted in decreased iron absorption, showing a competitive connection between the two elements. While excessive zinc levels can impair iron intake, zinc also serves as a cofactor for some enzymes involved in iron metabolism. According to studies, zinc is needed for the action of ferritin, an iron storage protein, and other enzymes that govern iron transport and usage in plants. Zinc biofortification techniques designed to increase zinc content in crops may have an indirect influence on iron concentrations. Involving zinc in enzymatic activity and general plant health can affect iron absorption and use efficiency. As a result, perfecting zinc biofortification methods may contribute to improving iron biofortification.

Effect of Zn Application on Other Minerals

High levels of Zn can hinder the ability of plants to uptake and use copper. Excess amounts of zinc may competitively obstruct Cu uptake, leading to lower amounts of copper in the tissues of crop plants. Similarly, zinc enhancement may alter the plant's manganese percentage. According to s certain research, elevated zinc levels can inhibit manganese absorption, potentially resulting in manganese shortages in plant tissues. While zinc and phosphorus are involved in different metabolic processes, zinc application can influence phosphorus availability in the soil. Improved nutrient uptake mechanisms, possibly due to enhanced root development, may indirectly affect phosphorus concentrations in plants. Zinc biofortification, when not carefully managed, might disrupt the balance with other essential minerals, including calcium. Zinc and calcium do not directly interact (Marschner, 2011), review emphasizes the indirect influence of zinc on overall nutrient balance, potentially impacting calcium uptake. The inclusive study of interactions between nutrients is necessary for the successful application of comprehensive biofortification techniques.

The soil pH and soil nutrient level affect the bioavailability of Zinc and other minerals. Soil pH influences the availability and absorption of certain nutrients. Cakmak and Kutman (2018) found that the effect of zinc administration on nutrient absorption in maize is connected to soil conditions, notably pH. Higher soil pH was linked to increased zinc availability, emphasizing the relevance of environmental parameters in biofortification efforts.

Therefore, keeping a balanced nutritional profile is critical for overall plant health. Zinc is simply one part of a complicated network of nutritional interactions. Zinc concentration changes may have a cascade effect on plants' entire nutritional balance. White and Broadley (2009) thorough study highlights the need to keep a balanced nutritional profile, highlighting the interconnection of multiple micronutrients, including zinc, for best plant development and health.

In conclusion, the effects of zinc biofortification on plant nutrient concentrations are many and controlled by a wide range of circumstances. While zinc is essential for plant growth, a thorough understanding of its interactions with other nutrients is needed to develop successful biofortification solutions. Future studies should go further into individual plant species, soil conditions, and agronomic approaches to fully realize the promise of zinc biofortification in addressing global nutritional concerns.

Microbial Application for Biofortification of Crops with Zinc Mobilization in Soil

Plants release chemicals from roots to promote interaction between microbes and plants as they adapt to metalliferous environments. Beneficial microorganisms (PGPMs) are vital for microbe-assisted phytoremediation because they lower metal phytotoxicity, enhance plant growth, and change metal bioavailability in soil (Ma et al., 2016). Zinc (Zn), which is required by plants at micro concentrations, has low bioavailability in soil due to a variety of causes. Zinc is absorbed from the soil solution by plant roots via membrane transport channels and stays as a divalent cation inside the plant. Transporter proteins such as P1B-ATPase, ZRT, IRT-like protein (ZIP), NRAMP, and CDF are needed for zinc absorption into cells. Organic compounds and zinc-solubilizing biological inoculations have proved potential to enhance plant zinc absorption (Hamzah Saleem et al., 2022). *Bacillus subtilis* (DS-178) and *Arthrobacter sp.* (DS-179) are identified as promising bacterial strains for zinc accumulation in grain (Singh et al., 2018).

Plant-Microbe Interactions in Zinc Uptake

Microbes and plants collaborated by creating several interaction mechanisms. However, it is critical to recognize the beneficial function that soil bacteria that promote plant development have in terms of plant productivity. Soil microorganisms also help plants grow and thrive under adverse conditions by regulating metabolic processes. Under unfavorable stress circumstances, plant growth-promoting microorganisms (PGPM) generate phytohormones, impart resistance against phytopathogens, and keep the right nutritional balance (Das et al., 2022). The clearest is the mycorrhizal response, which increases plant Zn accumulation when plants are cultivated in soil deficient in Zn. Through an indirect mechanism (inducing changes in root shape) or a direct pathway (extra radical hyphae), AMF boosted the uptake of Zn (Jansa et al., 2003). While the zinc content of roots or shoots rose when soil zinc concentration increased in both mycorrhizal and non-mycorrhizal plants, there was no clear difference in yield attributes (Audet & Charest, 2007).

Application of Microbial Biofortification

Utilizing biofertilizers is a cost-efficient, long-term solution to nutrient imbalance issues. Plants can access and preserve soil micronutrients with the aid of biofertilizers. One of the most serious issues is zinc deficiency because of its serious biological consequences and some concerning estimates provided by international organizations. The Food and Agriculture Organization (FAO) of the United Nations reports that zinc deficiency affects 50% of the world's cereal soils. The World Health Organization estimates that the health issues brought on by zinc deficiency claim the lives of about 800,000 people yearly, most of whom are children under five. Worldwide, around 2 billion people suffer from zinc insufficiency.

Localization, Speciation, and Bioavailability of Zinc in Agronomically Biofortified Crops

In the past 50–60 years, research has been conducted to improve the yield of cereals, but it has resulted in a substantial reduction in the concentrations of essential micronutrients. Decreases in the zinc content may be caused by the dilution effect (Shewry et al., 2016). Therefore, it is necessary to study the localization of these micronutrients and the physiological factors affecting their bioavailability.

Zinc Localization and Role in its Bioavailability

The aleurone layer and embryo (100 mg Zn kg^{-1}) are rich in zinc whereas endosperm (10 mg Zn kg^{-1}) is low in it, indicating Zn is predominantly localized in these parts of the grain (Cakmak & Kutman, 2018; Ozturk et al., 2006). White flour is mainly endosperm which is inherently low in zinc (5–10 mg Zn kg^{-1}) and is not enough to meet the Zn dietary requirement (Cakmak et al., 2010b). Furthermore, the grain contains high amounts of phytates, and its reduction leads to enhanced zinc bioavailability and intestinal absorption (Egli et al., 2004; Gibson et al., 2010). These phytates are localized in large concentrations in aleurone and embryo whereas endosperm is low in it (Prom-u-Thai et al., 2008). In embryo, zinc is concentrated in the form of Zn-phytate while in endosperm it is concentrated by forming a complex with Nitrogen/Sulphur. This could provide evidence for the high bioavailability of Zn in the endosperm and diminished bioavailability in aleurone and embryo as Zn-N/S ligands are bioavailable while Zn-phytate is less biologically available (Cheah et al., 2019). The studies also demonstrate the interaction between zinc and proteins in cereal grains, and therefore, grain proteins constitute a physiological sink for Zn (Ozturk et al., 2006; Persson et al., 2009; Cakmak et al., 2010b). Hence, it is necessary to enhance the zinc percentage in the endosperm part of the grain which is easily available for human consumption.

Agronomic Biofortification in Enhance Bioavailabe Zn

Zn is either absorbed by roots or deposited in leaves and stems and ultimately translocated into grains (Kutman et al., 2012; Sperotto, 2013). Approximately 70% of zinc is remobilized from leaves and stems into the grains, a crucial pathway for zinc accumulation (Grusak et al., 1999). The studies indicate that mobility of zinc is efficient through phloem and therefore, foliar application may be highly suitable for improving bioavailable Zn grain content. The remobilization/translocation of Zn is dependent on the availability of micronutrients and water at the grain-filling stage, and other factors. Recent wheat, rice, and maize experiments advocated that foliar sprays of zinc at the grain-filling stage were very efficacious in enhancing the cereal grain zinc content (up to 83%) as compared to the soil zinc fertilization at sowing (Cakmak et al., 2010a; Boonchuay et al., 2013; Abdoli et al., 2014). However, both Zn fertilization strategies improve zinc content specifically in the endosperm (white flour) (Cakmak et al., 2010a, b). In some studies, it has been suggested that the combination of soil fertilization and foliar application of micronutrients should be used to obtain better and more promising results (Szerement et al., 2022).

Environmental and Ecological Considerations

Zinc is an essential component for plant growth and development. It has a significant ecological influence on agriculture, changing various biological processes inside plants and, as consequently, impacting ecosystem health and productivity.

Enzyme Activation and Metabolism Regulation

Zinc activates enzymes in plant metabolic processes including photosynthesis, respiration, and DNA synthesis. Enzymes like carbonic anhydrase and alcohol dehydrogenase, which are essential for carbon fixation and energy metabolism in plants, require zinc as a cofactor (Cakmak & Marschner, 1988). Zinc is an intracellular regulator that offers structural support to proteins during molecular interactions, as well as maintaining the stability and integrity of biological membranes and ion channels. It serves as a structural element in nucleic acids and other gene-regulating proteins (Baltaci et al., 2018).

Protein Synthesis and Growth Regulation

Zinc promotes protein synthesis and affects gene expression in plant growth and development (Hacisalihoglu & Kochian, 2003). It is involved in the manufacturing of auxins, a kind of plant hormone that controls cell elongation and division, hence impacting total plant growth. Plant cells need a complex network of proteins to maintain appropriate intracellular concentration, which includes membrane transport proteins, cation transporters, cation diffusion facilitators, Zn-binding proteins, ligands, ZIP family members, and metal tolerance proteins. These proteins are preserved in terrestrial plants. They also coordinate processes such as sequestration, zinc absorption from the soil, and transport to ensure that all plant cells receive appropriate zinc (Khan et al., 2022).

Soil Fertility and Microbial Activity

Zinc affects microbial activity and nutrient cycling activities, which helps to maintain soil fertility (Kabata-Pendias, 2000). It functions as a cofactor for enzymes involved in organic matter breakdown, nitrogen fixation, and nutrient mineralization, hence improving soil health and production.

Economic Considerations and Practical Strategies

Efficacious adoption of Zn biofortification technology confronts numerous financial questions and adoption barriers. Hence, we will investigate the market concerns of the biofortification of wheat with zinc and propose pragmatic policies to surmount adoption barriers.

Economic Considerations

Cost-Benefit Ratio

Even though Zn Agronomic or genetic or other methods of biofortification of wheat is a lucrative and sustainable approach to regulate micronutrient deficiency (MND) (Sendhil et al., 2022), it increases the cost of production in terms of seed cost, and fertilizer application or microbe inoculum. This can be mitigated by the yield potential, market price, healthcare economic benefits, and Government subsidies and tax incentives. Therefore, a comprehensive cost-benefit ratio should be calculated, considering all these parameters, to estimate the economic expediency of the adoption of biofortified wheat with Zn. The fertilizer and foliar application of Zn or microbe inoculation has been reported to not just enhance Zn contents (99%) but also overall yield (Sendhil et al., 2022). Overall, this increases the economic returns for the farmers and saves costs in terms of healthcare.

Demand and Price

Price and demand are linked to one another and are an important component in analyzing the finance of biofortified wheat. Therefore, proper research and study of the market, consumer preferences and needs, and willingness to pay the high price for the product should be carried out. Malnutrition, a major global health issue, impacts about 1.5-2B people worldwide (Kumssa et al., 2015).

Storage and Distribution

Fortifying storage facilities, guaranteeing efficacious distribution of the seed, and cementing the supply chain will assist minimize costs and improve the adoption of the technology. Continuously striving to identify bottlenecks and subdue them is crucial in countering economic and adoption challenges.

Practical Strategies

To accelerate the Zn-biofortified wheat adoption it is necessary to execute practical methodologies that have an encouraging impression on human nutrition, health, and financial well-being.

Awareness and Education

It is important to educate and spread awareness among the malnourished communities, farmers, and policymakers, about the hidden hunger and its remedy and the importance of Zn-biofortification for farmers and malnourished people. To achieve it, conferences, stakeholder meetings, policy dialogues, and training programs should be organized to dissipate myths.

Demonstration

To encourage farmers to cultivate Zn-biofortified wheat at a large scale, various demonstration trials should be arranged at multiple locations. Each trial plot should be at least 1 acre with a local check in comparison to the Zn-biofortified line to get an accurate representation.

Technology Transfer and Capacity Building

The technology, i.e. seed and production manual will be transferred to the farmers, and they will be trained in agronomic practices. Moreover, sustainable seed distribution and extension channels should be developed to ensure the provision of high-quality seeds, Zn-fertilizers, Foliar spray, and agritech tools for the efficient adoption of Zn-rich varieties.

Farmer-Industry-Academia Linkages

Develop direct linkages between academic and research institutes, the seed industry, and farmers without a third party and such collaborations will further technology transfer, increase investment, and create a knowledge-sharing system. This will ultimately enhance the production of Zn-rich wheat varieties, ameliorate market access, and address challenges.

Inclusive Approach

Such an approach should be adopted to resolve the needs and problems of small farmers and malnourished marginalized rural communities by assisting with finances and resources.

Missing Pieces of the Puzzle and Future Perspectives

Identifying knowledge gaps and defining research objectives in zinc biofortification are essential for moving the field forward and addressing future challenges. Here are some potential areas of attention:

Understanding Zinc Bioavailability

While biofortification increases zinc content in crops, its bioavailability to humans is determined by a number of factors, including soil conditions, genotype, and food processing procedures (Hotz & Brown, 2004). Collaborative studies in agronomy, food science, and nutrition might reveal how these factors impact zinc bioavailability, resulting in more successful biofortification approaches.

Genetic Diversity and Trait Development

Current biofortification activities are mostly directed at staple crops such as maize, rice, and wheat (Kutman, 2010). However, investigating genetic variation in other crops and wild relatives may yield novel characteristics for increasing zinc concentration. Interdisciplinary cooperation across plant breeding, genetics, and bioinformatics might speed up the identification and development of high-zinc variants in various crops.

Climate Change Resilience

Climate change creates considerable hurdles for agricultural output, influencing soil nutrient availability and crop performance. Climate scientists, agronomists, and breeders might conduct interdisciplinary research to generate climate-resilient biofortified crop types that can thrive in changing climatic conditions (Knox et al., 2012).

Social and Behavioral Variables

The adoption of biofortified crops is influenced not only by their nutritional advantages but also by socio-cultural and economic considerations. Research that integrates social sciences, economics, and nutrition might uncover adoption hurdles and devise focused interventions to enhance the acceptability and consumption of biofortified foods (Meenakshi et al., 2012).

Conclusion

Zinc use through soil, foliar, and seed approaches is an achievable way of addressing zinc deficiency in crops and increasing their nutritional value for human consumption. Farmers may use these approaches to maximize zinc absorption by plants, resulting in increased crop yields and nutritional value. Furthermore, zinc-fortified biofortified foods can play an important role in fighting malnutrition, especially in areas where zinc deficiency is common. However, successful implementation requires thorough examination of multiple elements, including soil conditions, crop species, application techniques, and environmental implications. Agronomic biofortification is a viable method for improving zinc levels in a variety of crops, including cereals, legumes, vegetables, feed, and fodder crops. Researchers can increase zinc absorption and translocation inside plants by improving soil management methods, fertilizer administration strategies, and crop selection, resulting in higher nutritional quality. By utilizing microbial solutions, we can improve the efficiency and sustainability of zinc biofortification projects. Continued research and collaboration among scientists, politicians, and agricultural stakeholders are required to optimize the efficacy and sustainability of zinc biofortification solutions, contributing to worldwide efforts to combat hunger and enhance public health.

References

Abdoli, M., Esfandiari, E., Mousavi, S. B., & Sadeghzadeh, B. (2014). Effects of foliar application of zinc sulfate at different phenological stages on yield formation and grain zinc content of bread wheat (cv. Kohdasht). *Azarian Journal of Agriculture, 1*, 11–16.

Adnan, M., Tampubolon, K., ur Rehman, F., Saeed, M. S., Hayyat, M. S., Imran, M., Tahir, R., & Mehta, J. (2021). Influence of foliar application of magnesium on horticultural crops: A review. *Agrinula: Jurnal Agroteknologi Dan Perkebunan, 4*(1), 13–21.

Ali, I., Khan, A., Ali, A., Ullah, Z., Dai, D.-Q., Khan, N., Khan, A., Al-Tawaha, A. R., & Shei, H. (2022). Iron and zinc micronutrients and soil inoculation of Trichoderma harzianum enhance wheat grain quality and yield. *Frontiers in Plant Science, 13*, 960948.

Alloway, B. J. (2008). *Zinc in soils and crop nutrition.* http://www.topsoils.co.nz/wp-content/uploads/2014/09/Zinc-in-Soils-and-Crop-Nutrition-Brian-J.-Alloway.pdf

Alloway, B. J. (2009). Soil factors associated with zinc deficiency in crops and humans. *Environmental Geochemistry and Health, 31*(5), 537–548. https://doi.org/10.1007/s10653-009-9255-4

Almubarak, T., Ng, J. H., Ramanathan, R., & Nasr-El-Din, H. A. (2021). Chelating agents for oilfield stimulation: Lessons learned and future outlook. *Journal of Petroleum Science and Engineering, 205*, 108832. https://doi.org/10.1016/j.petrol.2021.108832

Anwar, Z., Basharat, Z., Bilal Hafeez, M., Zahra, N., Rafique, Z., & Maqsood, M. (2021). Biofortification of maize with zinc and iron not only enhances crop growth but also improves grain quality. *Asian Journal of Agriculture and Biology.* Online. http://research.manuscritpub.com/id/eprint/1950/

Arts, J., Bade, S., Badrinas, M., Ball, N., & Hindle, S. (2018). Should DTPA, an Aminocarboxylic acid (ethylenediamine-based) chelating agent, be considered a developmental toxicant? *Regulatory Toxicology and Pharmacology, 97*, 197–208. https://doi.org/10.1016/j.yrtph.2018.06.019

Athar, T., Khan, M. K., Pandey, A., Yilmaz, F. G., Hamurcu, M., Hakki, E. E., & Gezgin, S. (2020). Biofortification and the involved modern approaches. *Journal of Elementology, 25*(2). https://agro.icm.edu.pl/agro/element/bwmeta1.element.agro-c8945cfc-c28e-4932-ba14-a0c9dd399d83

Audet, P., & Charest, C. (2007). Dynamics of arbuscular 1 mycorrhizal symbiosis in heavy metal phytoremediation: Meta-analytical and conceptual perspectives. *Environmental Pollution, 147*)(3), 609–614.

Baltaci, A. K., Yuce, K., & Mogulkoc, R. (2018). Zinc metabolism and metallothioneins. *Biological Trace Element Research, 183*(1), 22–31. https://doi.org/10.1007/s12011-017-1119-7

Bana, R. S., Jat, G. S., Grover, M., Bamboriya, S. D., Singh, D., Bansal, R., Choudhary, A. K., Kumar, V., Laing, A. M., & Godara, S. (2022). Foliar nutrient supplementation with micronutrient-embedded fertilizer increases biofortification, soil biological activity and productivity of eggplant. *Scientific Reports, 12*(1), 5146.

Bhantana, P., Moussa, M. G., Malla, R., Khadka, D., Vista, S. P., Shrestha, R. K., & Hu, C. X. (2022). Foliar versus soil biofortification of Zn in citrus (citrus reticulata Blanco) effect on mineral nutrition and fruit yield and quality. *Biomedical Journal of Scientific & Technical Research, 41*, 32755–32568.

Bibi, F., Saleem, I. S., Ehsan, S., Jamil, S., Ullah, H., Mubashir, M., Kiran, S., Ahmad, I., Irshad, I., & Saleem, M. (2020). Effect of various application rates of phosphorus combined with different zinc rates and time of zinc application on phytic acid concentration and zinc bioavailability in wheat. *Agriculture and Natural Resources, 54*(3), 265–272.

Boonchuay, P., Cakmak, I., Rerkasem, B., & Prom-U-Thai, C. (2013). Effect of different growth stages on seed zinc concentration and its impact on seedling vigor in rice. *Journal of Soil Science & Plant Nutrition, 59*, 180–188.

Brown, K. H., Wuehler, S. E., & Peerson, J. M. (2001). The importance of zinc in human nutrition and estimation of the global prevalence of zinc deficiency. *Food and Nutrition Bulletin, 22*(2), 113–125. https://doi.org/10.1177/156482650102200201

Cakmak, I. (2002). Plant nutrition research: Priorities to meet human needs for food in sustainable ways. *Plant and Soil, 247*, 3–24.

Cakmak, I. (2008). *Plant nutrition research: Priorities to meet human needs for food in sustainable ways*.

Cakmak, I., & Kutman, U. B. (2018). Agronomic biofortification of cereals with zinc: A review. *European Journal of Soil Science, 69*(1), Article 1. https://doi.org/10.1111/ejss.12437

Cakmak, I., & Marschner, H. (1988). Zinc-dependent changes in ESR signals, NADPH oxidase and plasma membrane permeability in cotton roots. *Physiologia Plantarum, 73*(1), 182–186. https://doi.org/10.1111/j.1399-3054.1988.tb09214.x

Cakmak, I., Kalayci, M., Kaya, Y., Torun, A. A., Aydin, N., Wang, Y., et al. (2010a). Biofortification and localization of zinc in wheat grain. *Journal of Agricultural & Food Chemistry, 58*, 9092–9102.

Cakmak, I., Pfeiffer, W. H., & McClafferty, B. (2010b). Biofortification of durum wheat with zinc and iron. *Cereal Chemistry, 87*, 10–20.

Cheah, Z. X., Kopittke, P. M., Harper, S. M., Meyer, G., O'Hare, T. J., & Bell, M. J. (2019). Speciation and accumulation of Zn in sweetcorn kernels for genetic and agronomic biofortification programs. *Planta, 250*(1), 219–227. https://doi.org/10.1007/s00425-019-03162-x

Das, P. P., Singh, K. R., Nagpure, G., Mansoori, A., Singh, R. P., Ghazi, I. A., Kumar, A., & Singh, J. (2022). Plant-soil-microbes: A tripartite interaction for nutrient acquisition and better plant growth for sustainable agricultural practices. *Environmental Research, 214*, 113821. https://doi.org/10.1016/j.envres.2022.113821

Dimka Haytova, D. (2013). A review of foliar fertilization of some vegetables crops. *Annual Research & Review in Biology, 3*(4), 455–465.

dos Santos, T. B., Ribas, A. F., de Souza, S. G. H., Budzinski, I. G. F., & Domingues, D. S. (2022). Physiological responses to drought, salinity, and heat stress in plants: A review. *Stresses, 2*(1), 113–135.

Egli, I., Davidsson, L., Zeder, C., Walczyk, T., & Hurrell, R. (2004). Dephytinization of a complementary food based on wheat and soy increases zinc, but not copper, apparent absorption in adults. *Journal of Nutrition, 134*, 1077–1080.

Eifediyi, E. K., Ilori, G. A., Ahamefule, H. E., & Imam, A. Y. (2021). The effects of zinc biofortification of seeds and NPK fertilizer application on the growth and yield of sesame (Sesamum indicum L.). *Acta Agriculturae Slovenica, 117*(1), Article 1. https://doi.org/10.14720/aas.2021.117.1.1252

Gao, X., Zhang, L., Peng, Y., Ding, J., & An, Z. (2023). The successful integration of anammox to enhance the operational stability and nitrogen removal efficiency during municipal wastewater treatment. *Chemical Engineering Journal, 451*, 138878. https://doi.org/10.1016/j.cej.2022.138878

Gibson, R. S. (2006). Zinc: The missing link in combating micronutrient malnutrition in developing countries. *Proceedings of the Nutrition Society, 65*(1), 51–60.

Gibson, R. S., Bailey, K. B., Gibbs, M., & Ferguson, E. L. (2010). A review of phytate, iron, zinc and calcium concentrations in plant-based complementary foods used in low-income countries and implications for bioavailability. *Food & Nutrition Bulletin, 31*, S134–S146.

Grusak, M. A., Pearson, J. N., & Marentes, E. (1999). The physiology of micronutrient homeostasis in field crops. *Field Crops Research, 60*, 41–56.

Hacisalihoglu, G., & Kochian, L. V. (2003). How do some plants tolerate low levels of soil zinc? Mechanisms of zinc efficiency in crop plants. *New Phytologist, 159*(2), 341–350. https://doi.org/10.1046/j.1469-8137.2003.00826.x

Hamzah Saleem, M., Usman, K., Rizwan, M., Al Jabri, H., & Alsafran, M. (2022). Functions and strategies for enhancing zinc availability in plants for sustainable agriculture. *Frontiers in Plant Science, 13*, 1033092. https://doi.org/10.3389/fpls.2022.1033092

Hassan, N. S., Din, T. A. S. E., Hendawey, M. H., Borai, I. H., & Mahdi, A. A. (2024). Magnetite and zinc oxide nanoparticles alleviated heat stress in wheat plants. *Current Nanomaterials, 3*(1), Article 1.

Hong, J., Wang, C., Wagner, D. C., Gardea-Torresdey, J. L., He, F., & Rico, C. M. (2021). Foliar application of nanoparticles: Mechanisms of absorption, transfer, and multiple impacts. *Environmental Science: Nano, 8*(5), 1196–1210. https://doi.org/10.1039/D0EN01129K

Hotz, C., & Brown, K. H. (2004). *Assessment of the risk of zinc deficiency in populations and options for its control*.

Hussain, S., Maqsood, M. A., Rengel, Z., & Aziz, T. (2012). Biofortification and estimated human bioavailability of zinc in wheat grains as influenced by methods of zinc application. *Plant and Soil, 361*, 279–290.

Impa, S. M., & Johnson-Beebout, S. E. (2012). Mitigating zinc deficiency and achieving high grain Zn in rice through integration of soil chemistry and plant physiology research. *Plant and Soil, 361*, 3–41.

Jansa, J., Mozafar, A. & Frossard, E. (2003). Long-distance transport of P and Zn through the hyphae of an arbuscular mycorrhizal fungus in symbiosis with maize. *Agronomie, 23*(5–6), 481-488.

Kabata-Pendias, A. (2000). *Trace elements in soils and plants*. CRC Press. https://www.taylorfrancis.com/books/mono/10.1201/9781420039900/trace-elements-soils-plants-alina-kabata-pendias

Kadi, V. P., Vishwavidyalaya, S., Rudani, K., Patel, V., & Prajapati, K. (2018). The importance of zinc in plant growth—A review. *International Research Journal of Applied Sciences, 46*, 2349–4077.

Kaur, A., & Singh, G. (2022). Zinc and iron application in conjunction with nitrogen for agronomic biofortification of field crops – A review. *Crop and Pasture Science*. https://doi.org/10.1071/CP21487

Khan, S. T., Malik, A., & Ahmad, F. (2022). Role of zinc homeostasis in plant growth. In S. T. Khan & A. Malik (Eds.), *Microbial biofertilizers and micronutrient availability: The role of zinc in agriculture and human health* (pp. 179–195). Springer International Publishing. https://doi.org/10.1007/978-3-030-76609-2_9

Knox, J., Hess, T., Daccache, A., & Wheeler, T. (2012). Climate change impacts on crop productivity in Africa and South Asia. *Environmental Research Letters, 7*(3), 034032.

Krishnasree, R. K., Raj, S. K., & Chacko, S. R. (2021). Foliar nutrition in vegetables: A review. *Journal of Pharmacognosy and Phytochemistry, 10*(1), 2393–2398.

Kumssa, D. B., Joy, E. J. M., Ander, E. L., Watts, M. J., Young, S. D., Walker, S., & Broadley, M. R. (2015). Dietary calcium and zinc deficiency risks are decreasing but remain prevalent. *Scientific Reports, 5*, 10974. https://doi.org/10.1038/srep109745

Kutman, B. Ü. (2010). *Roles of nitrogen and zinc nutrition in biofortification of wheat grain* (Ph.D. thesis). https://research.sabanciuniv.edu/id/eprint/24073/

Kutman, U. B., Kutman, B. Y., Ceylan, Y., Ova, E. A., & Cakmak, I. (2012). Contributions of root uptake and remobilization to grain zinc accumulation in wheat depending on post-anthesis zinc availability and nitrogen nutrition. *Plant and Soil, 361*, 177–187.

Li, J., Lens, P. N. L., Otero-Gonzalez, L., & Du Laing, G. (2020). Production of selenium- and zinc-enriched *Lemna* and *Azolla* as potential micronutrient-enriched bioproducts. *Water Research, 172*, 115522. https://doi.org/10.1016/j.watres.2020.115522

Liu, D.-Y., Liu, Y.-M., Zhang, W., Chen, X.-P., & Zou, C.-Q. (2019). Zinc uptake, translocation, and remobilization in winter wheat as affected by soil application of Zn fertilizer. *Frontiers in Plant Science, 10*, 443999.

Liu, L., Melse-Boonstra, A., Van Der Werf, W., Zhang, F., Cong, W., & Stomph, T. J. (2024). The potential of biofortification technologies for wheat and rice to fill the nutritional Zn intake gap in China. *Journal of the Science of Food and Agriculture, 104*(5), 2651–2659. https://doi.org/10.1002/jsfa.13149

Ma, Y., Oliveira, R. S., Freitas, H., & Zhang, C. (2016). Biochemical and molecular mechanisms of plant-microbe-metal interactions: Relevance for phytoremediation. *Frontiers in Plant Science, 7*, 918.

Maqbool, M. A., & Beshir, A. (2019). Zinc biofortification of maize (Zea mays L.): Status and challenges. *Plant Breeding, 138*(1), 1–28. https://doi.org/10.1111/pbr.12658

Marschner, H. (2011). *Marschner's mineral nutrition of higher plants*. Academic.

Meenakshi, J. V., Banerji, A., Manyong, V., Tomlins, K., Mittal, N., & Hamukwala, P. (2012). Using a discrete choice experiment to elicit the demand for a nutritious food: Willingness-to-pay for orange maize in rural Zambia. *Journal of Health Economics, 31*(1), 62–71.

Mehboob, U., Sarwar, G., & Tahir, M. A. (2022). Role of iron biofortification to improve growth, yield and chemical composition of maize. *Journal of Pure and Applied Agriculture, 7*(2). https://ojs.aiou.edu.pk/index.php/jpaa/article/download/991/860

Mengel, K., Kirkby, E. A., Kosegarten, H., & Appel, T. (Eds.). (2001). *Principles of plant nutrition*. Springer Netherlands. https://doi.org/10.1007/978-94-010-1009-2

Montalvo, D., Degryse, F., da Silva, R. C., Baird, R., & McLaughlin, M. J. (2016). Chapter five—Agronomic effectiveness of zinc sources as micronutrient fertilizer. In D. L. Sparks (Ed.), *Advances in agronomy* (Vol. 139, pp. 215–267). Academic. https://doi.org/10.1016/bs.agron.2016.05.004

Mostofa, M. G., Rahman, M. M., Ghosh, T. K., Kabir, A. H., Abdelrahman, M., Rahman Khan, M. A., Mochida, K., & Tran, L.-S. P. (2022). Potassium in plant physiological adaptation to abiotic stresses. *Plant Physiology and Biochemistry, 186*, 279–289. https://doi.org/10.1016/j.plaphy.2022.07.011

National Nutrition Survey 2018 | UNICEF Pakistan. (n.d.). Retrieved April 4, 2024, from https://www.unicef.org/pakistan/national-nutrition-survey-2018

Niu, J., Liu, C., Huang, M., Liu, K., & Yan, D. (2021). Effects of foliar fertilization: A review of current status and future perspectives. *Journal of Soil Science and Plant Nutrition, 21*(1), 104–118. https://doi.org/10.1007/s42729-020-00346-3

Ozturk, L., Yazici, M. A., Yucel, C., Torun, A., Cekic, C., Bagci, A., et al. (2006). Concentration and localization of zinc during seed development and germination in wheat. *Physiologia Plantarum, 128*, 144–152.

Persson, D. P., Hansen, T. H., Laursen, K. H., Schjoerring, J. K., & Husted, S. (2009). Simultaneous iron, zinc, sulfur and phosphorus speciation analysis of barley grain tissues using SEC-ICP-MS and IP-ICP-MS. *Metallomics, 1*, 418–426.

Phattarakul, N., Rerkasem, B., Li, L. J., Wu, L. H., Zou, C. Q., Ram, H., Sohu, V. S., Kang, B. S., Surek, H., Kalayci, M., Yazici, A., Zhang, F. S., & Cakmak, I. (2012). Biofortification of rice grain with zinc through zinc fertilization in different countries. *Plant and Soil, 361*(1), 131–141. https://doi.org/10.1007/s11104-012-1211-x

Praharaj, S., Skalicky, M., Maitra, S., Bhadra, P., Shankar, T., Brestic, M., Hejnak, V., Vachova, P., & Hossain, A. (2021). Zinc biofortification in food crops could alleviate the zinc malnutrition in human health. *Molecules, 26*(12), Article 12. https://doi.org/10.3390/molecules26123509

Prasad, R., Shivay, Y. S., & Kumar, D. (2014). Chapter two—Agronomic biofortification of cereal grains with iron and zinc. In D. L. Sparks (Ed.), *Advances in agronomy* (Vol. 125, pp. 55–91). Academic. https://doi.org/10.1016/B978-0-12-800137-0.00002-9

Prom-u-Thai, C., Huang, L., Rerkasem, B., Thomson, G., Kuo, J., Saunders, M., et al. (2008). Distribution of protein bodies and phytate-rich inclusions in grain tissues of low and high iron rice genotypes. *Cereal Chemistry, 85*, 257–265.

Rakotondramanana, M., Wissuwa, M., Stangoulis, J., & Grenier, C. (2024). Stability of grain zinc concentrations across lowland rice environments favors zinc biofortification breeding. *Frontiers in Plant Science, 15*, 1293831.

Rayman, M. P. (2012). Selenium and human health. *Lancet (London, England), 379*(9822), Article 9822. https://doi.org/10.1016/S0140-6736(11)61452-9

Raza, M. A., Saeed, A., Munir, H., Ziaf, K., Shakeel, A., Saeed, N., Munawar, A., & Rehman, F. (2017). Screening of tomato genotypes for salinity tolerance based on early growth attributes and leaf inorganic osmolytes. *Archives of Agronomy and Soil Science, 63*(4), 501–512. https://doi.org/10.1080/03650340.2016.1224856

Rengel, Z. (2001). Genotypic differences in micronutrient use efficiency in crops. *Communications in Soil Science and Plant Analysis, 32*(7–8), 1163–1186.

Rengel, Z., Batten, G. D., & Crowley, D. E. (1999). Agronomic approaches for improving the micronutrient density in edible portions of field crops. *Field Crops Research., 60*, 27–40.

Reshma, Z., & Meenal, K. (2024). Zinc biofortification and implications on growth, nutrient efficiency, and stress response in Amaranthus cruentus through soil application of biosynthesized nanoparticles. *Applied Food Research, 4*(1), 100385.

Rizwan, M., Dalimunthe, M., Pasaribu, I. A., & Satriawan, H. (2021). The effect of organic fertilizers on growth several varieties of soybeans. *IOP Conference Series: Earth and Environmental Science, 883*(1), Article 1. https://doi.org/10.1088/1755-1315/883/1/012051

Sendhil, R., Adeeth Cariappa, A. G., Ramasundaram, P., Vikas, G., Gopalareddy, K., Gupta, O. P., Anuj, K., Satyavir, S., & Singh, G. P. (2022). Biofortification in wheat: Research progress, potential impact, and policy imperatives. *AgriRxiv, 1*.

Shahane, A. A., & Shivay, Y. S. (2022). Agronomic biofortification of crops: Current research status and future needs. *Indian Journal of Fertilisers, 18*(2), 164–179.

Shewry, P. R., Pellny, T. K., & Lovegrove, A. (2016). Is modern wheat bad for health? *Nature Plants, 2*, 1–3.

Singh, D., Geat, N., & Rajawat, M.V.S. et al. (2018). Prospecting endophytes from different Fe or Zn accumulating wheat genotypes for their influence as inoculants on plant growth, yield, and micronutrient content. *Annals of Microbiology, 68*, 815–833.

Singh Dhaliwal, S., Sharma, V., Kumar Shukla, A., Singh Shivay, Y., Hossain, A., Verma, V., Kaur Gill, M., Singh, J., Singh Bhatti, S., Verma, G., Singh, J., & Singh, P. (2023). Agronomic biofortification of forage crops with zinc and copper for enhancing nutritive potential: A systematic review. *Journal of the Science of Food and Agriculture, 103*(4), 1631–1643. https://doi.org/10.1002/jsfa.12353

Smith, J., Grimmer, M., Waterhouse, S., & Paveley, N. (2013). Quantifying the non-fungicidal effects of foliar applications of fluxapyroxad (Xemium) on stomatal conductance, water use efficiency and yield in winter wheat. *Communications in Agricultural and Applied Biological Sciences, 78*(3), 523–535.

Sperotto, R. A. (2013). Zn/Fe remobilization from vegetative tissues to rice seeds: Should I stay or should I go? Ask Zn/Fe supply. *Frontiers in Plant Science, 4*, 464.

Stanton, C., Sanders, D., Krämer, U., & Podar, D. (2022). Zinc in plants: Integrating homeostasis and biofortification. *Molecular Plant, 15*(1), 65–85.

Stomph, T. J., Jiang, W., & Struik, P. C. (2009). Zinc biofortification of cereals: Rice differs from wheat and barley. *Trends in Plant Science, 14*(3), Article 3. https://doi.org/10.1016/j.tplants.2009.01.001

Suganya, A., Saravanan, A., & Manivannan, N. (2020). Role of zinc nutrition for increasing zinc availability, uptake, yield, and quality of maize (Zea mays L.) grains: An overview. *Communications in Soil Science and Plant Analysis, 51*(15), 2001–2021.

Szerement, J., Szatanik-Kloc, A., Mokrzycki, J., & Mierzwa-Hersztek, M. (2022). Agronomic biofortification with Se, Zn, and Fe: An effective strategy to enhance crop nutritional quality and stress defense-a review. *Journal of Soil Science and Plant Nutrition, 22*(1). Springer International Publishing. https://doi.org/10.1007/s42729-021-00719-2

Ullah, A., Nadeem, F., Nawaz, A., Siddique, K. H. M., & Farooq, M. (2022). Heat stress effects on the reproductive physiology and yield of wheat. *Journal of Agronomy and Crop Science, 208*(1), 1–17. https://doi.org/10.1111/jac.12572

White, P. J., & Broadley, M. R. (2009). Biofortification of crops with seven mineral elements often lacking in human diets—Iron, zinc, copper, calcium, magnesium, selenium and iodine. *The New Phytologist, 182*(1), Article 1. https://doi.org/10.1111/j.1469-8137.2008.02738.x

Widodo, B., Broadley, M. R., Rose, T., Frei, M., Pariasca-Tanaka, J., Yoshihashi, T., Thomson, M., Hammond, J. P., Aprile, A., Close, T. J., Ismail, A. M., & Wissuwa, M. (2010). Response to zinc deficiency of two rice lines with contrasting tolerance is determined by root growth maintenance and organic acid exudation rates, and not by zinc-transporter activity. *The New Phytologist, 186*(2), Article 2. https://doi.org/10.1111/j.1469-8137.2009.03177.x

Yeboah, S., Asibuo, J., Oteng-Darko, P., Asamoah Adjei, E., Lamptey, M., Owusu Danquah, E., Waswa, B., & Butare, L. (2021). Impact of foliar application of zinc and magnesium amino-chelate on bean physiology and productivity in Ghana. *International Journal of Agronomy, 2021*, 1–9.

Zhang, X., Zhai, H., Wang, Y., Tian, X., Zhang, Y., Wu, H., Lü, S., Yang, G., Li, Y., & Wang, L. (2016). Functional conservation and diversification of the soybean maturity gene E1 and its homologs in legumes. *Scientific Reports, 6*(1), 1–14.

Zulfiqar, U., Maqsood, M., & Hussain, S. (2020). Biofortification of rice with iron and zinc: Progress and prospects. In A. Roychoudhury (Ed.), *Rice research for quality improvement: Genomics and genetic engineering* (pp. 605–627). https://doi.org/10.1007/978-981-15-5337-0_26

Zulfiqar, U., Hussain, S., Ishfaq, M., Ali, N., Yasin, M. U., & Ali, M. A. (2021). Foliar manganese supply enhances crop productivity, net benefits, and grain manganese accumulation in direct-seeded and puddled transplanted rice. *Journal of Plant Growth Regulation, 40*(4), 1539–1556.

Open Access This chapter is licensed under the terms of the Creative Commons Attribution 4.0 International License (http://creativecommons.org/licenses/by/4.0/), which permits use, sharing, adaptation, distribution and reproduction in any medium or format, as long as you give appropriate credit to the original author(s) and the source, provide a link to the Creative Commons license and indicate if changes were made.

The images or other third party material in this chapter are included in the chapter's Creative Commons license, unless indicated otherwise in a credit line to the material. If material is not included in the chapter's Creative Commons license and your intended use is not permitted by statutory regulation or exceeds the permitted use, you will need to obtain permission directly from the copyright holder.

The manufacturer's authorised representative in the EU is Springer Nature Customer Service Centre GmbH, Europaplatz 3, 69115 Heidelberg, Germany. If you have any concerns regarding our products, please contact ProductSafety@springernature.com

Printed and bound by CPI Group (UK) Ltd, Croydon, CR0 4YY

26/03/2026

02078974-0001